UNIVERSITY ENGAGEMENT WITH FARMING COMMUNITIES IN AFRICA

This book explains and explores how collaborations can be built and strengthened between African universities and farming communities to address real-world contemporary challenges.

The book focuses on Community Action Research Platforms, an approach that has successfully enabled African universities to break free of the ivory tower and prove their relevance to society through deep collaborative engagements in targeted agricultural value chains. Developed in a pan-African network of universities (RUFORUM) focused on capacity building in agriculture, the approach has been tested in diverse settings over the last 15 years. The book draws on the experiences and lessons from 21 different projects initiated by RUFORUM member universities in Benin, Botswana, Ethiopia, Ghana, Kenya, Malawi, Namibia, South Africa, Sudan, Tanzania, Uganda and Zimbabwe. It highlights a critical yet underutilised role for African universities as collaborators and catalysts for multi-sector solutions. These are solutions that increase productivity and address climate change. They develop livelihoods and resilience in rural communities, as well as promote farmers' access to markets, innovation and trade while safeguarding biodiversity and enhancing food and nutrition security. The book makes a case for repositioning African universities as fulcrums of development in society. It shares the rich experiences, learnings and scientific findings of diverse researchers, practitioners and students who have been working towards achieving this reality on the ground.

This multidisciplinary book holds appeal for university leaders, higher education, agrifood and development specialists, researchers and practitioners, policymakers and development agencies engaged in African agriculture and rural development, higher education and sustainable growth.

Anthony Egeru is the Training and Community Development Manager at the Regional Universities Forum for Capacity Building in Agriculture (RUFORUM) Secretariat and is an Associate Professor at Makerere University, Uganda. His research focuses on stochastic environmental change processes and social systems, and he has facilitated community action research across 23 African countries.

Megan Lindow is a writer based in Cape Town, South Africa, whose work focuses on documenting complex processes of development and change in spaces of education, research, agriculture and food systems, climate change, gender equality and sustainable development across Africa and beyond. She holds an MA in journalism from Stanford University, USA, and an MPhil in Sustainable Development from Stellenbosch University, South Africa.

Kay Muir Leresche is a former Professor of Agricultural and Natural Resource Economics at the University of Zimbabwe, Africa. After retiring, she chaired the RUFORUM Technical Committee from 2004 to 2017 and worked with many international development and research agencies. She is committed to supporting universities to catalyse the transformation of sustainable and inclusive agriculture in Africa.

Earthscan Food and Agriculture

Sustainable Apple Breeding and Cultivation in Germany
Commons-Based Agriculture and Social-Ecological Resilience
Hendrik Wolter

Food Policy in the United Kingdom
An Introduction
Martin Caraher, Sinéad Furey and Rebecca Wells

Peasants, Capitalism, and the Work of Eric R. Wolf
Reviving Critical Agrarian Studies
Mark Tilzey, Fraser Sugden and David Seddon

Peasants, Capitalism, and Imperialism in an Age of Politico-Ecological Crisis
Mark Tilzey and Fraser Sugden

Genome Editing and Global Food Security
Molecular Engineering Technologies for Sustainable Agriculture
Zeba Khan, Durre Shahwar and Yasmin Heikal

University Engagement with Farming Communities in Africa
Community Action Research Platforms
Edited by Anthony Egeru, Megan Lindow and Kay Muir Leresche

For more information about this series, please visit: www.routledge.com/books/series/ECEFA/

UNIVERSITY ENGAGEMENT WITH FARMING COMMUNITIES IN AFRICA

Community Action Research Platforms

Edited by
Anthony Egeru, Megan Lindow and Kay Muir Leresche

Routledge
Taylor & Francis Group
LONDON AND NEW YORK

earthscan
from Routledge

RUFORUM
Capacity Building in Agriculture

Designed cover image: © RUFORUM

First published 2024
by Routledge
4 Park Square, Milton Park, Abingdon, Oxon OX14 4RN

and by Routledge
605 Third Avenue, New York, NY 10158

Routledge is an imprint of the Taylor & Francis Group, an informa business

Funded by Regional Universities Forum for Capacity Building in Agriculture (RUFORUM).

British Library Cataloguing-in-Publication Data
A catalogue record for this book is available from the British Library

Library of Congress Cataloging-in-Publication Data
Names: Egeru, Anthony, editor. | Lindow, Megan, editor. | Muir-Leresche, Kay, editor.
Title: University engagement with farming communities in Africa : community action research platforms / edited by Anthony Egeru, Megan Lindow, and Kay Muir Leresche.
Description: New York : Routledge, 2024. | Includes bibliographical references and index.
Identifiers: LCCN 2023033400 (print) | LCCN 2023033401 (ebook) |
ISBN 9781032481203 (hardback) | ISBN 9781032481180 (paperback) |
ISBN 9781003387497 (ebook)
Subjects: LCSH: Universities and colleges—Africa, Sub-Saharan—Public services—Case studies. | Rural development projects—Africa, Sub-Saharan—Case studies. | Community-based research—Africa, Sub-Saharan—Case studies. | Community and college—Africa, Sub-Saharan—Case studies. | Agriculture—Research—Africa, Sub-Saharan—Case studies.
Classification: LCC LB2331.46.A357 U658 2024 (print) | LCC LB2331.46.A357 (ebook) |
DDC 378.1030967—dc23
LC record available at https://lccn.loc.gov/2023033400
LC ebook record available at https://lccn.loc.gov/2023033401

ISBN: 978-1-032-48120-3 (hbk)
ISBN: 978-1-032-48118-0 (pbk)
ISBN: 978-1-003-38749-7 (ebk)

DOI: 10.4324/9781003387497

Typeset in Times New Roman
by codeMantra

CONTENTS

FOREWORD

If there are two realities that will shape the future of Africa, it's these: one, agriculture is among the largest economic sectors of the continent – and Africa's single biggest employer. Second: Africa is the youngest continent in the world, with over 70 percent of the population under the age of thirty. Combined, these twin realities have an impact on almost all facets of Africa's future. They have implications for addressing and mitigating climate change. Reducing food insecurity. Bolstering health. In fact, it's impossible to have a conversation about Africa's socio-economic future apart from these two realities.

So, at the Mastercard Foundation, both these truisms are central to our work. In 2018, we launched our Young Africa Works strategy and set a bold target of enabling dignified and fulfilling work for 30 million young people, with a strong focus on the economic inclusion of young women, forcibly displaced youth and young people living with disabilities. The agricultural sector has pride of place in our strategy as one of the cross-cutting sectors that African countries have a stake in, and where there is a huge opportunity to deliver improvements in income and create more work opportunities.

We do so in part by getting young people engaged in innovating to address challenges facing the sector – and their communities. This book is a testament to what's possible when you train young people, connect them to communities and empower them to solve problems. It also demonstrates the critical role higher education institutions must play in that process.

Ultimately, this literature is the product of an ongoing, eight-year partnership between the Mastercard Foundation and RUFORUM, which we are incredibly proud of. We launched our partnership with RUFORUM in 2016, with the goal of enabling African agricultural universities and the young people they train to create solutions for the agricultural sector by testing, applying and scaling cutting-edge science, technology and agricultural practices, while bringing entrepreneurial energy into the sector. As part of the partnership, young people studying at RUFORUM universities were placed in community settings to problem-solve alongside community members on agricultural issues they were facing.

There is so much to learn from what young people have achieved through their community-based projects and engagements – particularly for institutions and organisations, like RUFORUM and the Foundation, that believe that developing solutions to pressing global challenges must start by engaging, equipping and enabling the next generation of leaders. Congratulations to RUFORUM for leading the way.

Reeta Roy
President and CEO, Mastercard Foundation

ACKNOWLEDGEMENTS

The projects, innovations, graduates, businesses and the personal growth of all the individuals and institutions engaged in the Community Action Research Platforms (CARPs) would not have been possible without the engagement and financial support given by the Mastercard Foundation. Their support to strengthen the relevance of higher education and their goal to leave no one behind have provided impetus and reach to RUFORUM, the Regional Universities Forum for Capacity Building in Africa. The RUFORUM Secretariat and the RUFORUM Technical Committee have worked tirelessly with the Principal Investigators of the CARPs to ensure their success and to overcome constraints. The authors would like to thank, in particular, Professor Adipala Ekwamu, the former RUFORUM Executive Secretary and Dr Patrick Okori the current Executive Secretary. Their vision and commitment have made this all possible.

The authors would also like to acknowledge all of the RUFORUM member universities that supported the CARPs, as well as all government departments and ministries, international agencies, TVETs, non-governmental organisations, private sector, research institutes, researchers and students who collaborated in the platforms.

The welcome, support and expertise brought by the smallholder farmers and their communities ensured the vibrancy and sustainability of the relationships formed.

CONTRIBUTORS

Ruhama Abdallh is with the Faculty of Natural Resources and Environmental Studies, Peace University in Sudan.

E. E. A. Ahmed is with the Institute of Desert & Desertification Studies, University of Khartoum in Sudan.

Tafaol E. Ali is with the Faculty of Agricultural Sciences, University of Gezira, Sudan.

Tony Aliro is a Lecturer in Veterinary Epidemiology at Gulu University in Uganda.

Simon Angombe is Associate Dean, School of Agriculture and Fisheries Sciences, University of Namibia.

Festus Annor-Frempong is an Associate Professor in Agricultural Extension Education at the University of Cape Coast in Ghana.

Peter F. Arama is from Rongo University in Kenya.

Achille E. Assogbadjo is Professor, Laboratory of Applied Ecology, Faculty of Agronomic Sciences, University of Abomey-Calavi in Benin.

Abel Atukwase is with the School of Food Technology, Nutrition and Bioengineering at Makerere University in Uganda.

Napoléon Heri Bahati-Kajunju is from the School of Food Technology, Nutrition and Bioengineering at Makerere University in Uganda.

Bockline O. Bebe is a Professor of Livestock Production Systems at Egerton University in Kenya.

Flora J. Chadaré is Professor, Laboratoire de Sciences et Technologie des Aliments et Bioressources et de Nutrition Humaine, École des Sciences et Techniques de Conservation et de Transformation des Produits Agricoles (ESTCTPA), Université Nationale d'Agriculture and the Laboratory of Food Sciences, Faculty of Agronomic Sciences, University of Abomey-Calavi in Benin.

Anthony Egeru is Training and Community Development Manager for the RUFORUM Secretariat and a Senior Lecturer at Makerere University in Kampala, Uganda.

Mutasim A. M. Elagab is with the Faculty of Agricultural Science, University of Gezira, Sudan.

Muna Mohamed Elhag is with the Water Management and Irrigation Institute, University of Gezira, Sudan.

Ali E. Eljack is with the Faculty of Agricultural Science, University of Gezira, Sudan.

Elhadi. A. I. Elkhalil is with the Department of Botany & Agricultural Biotechnology, Faculty of Agriculture, University of Khartoum in Sudan.

A. E. M. Elzein is with the Department of Botany & Agricultural Biotechnology, Faculty of Agriculture, University of Khartoum in Sudan.

Vallantino Emongor is a Professor of Horticulture at Botswana University of Agriculture and Natural Resources.

Rodrigue C. Gbèdomon is with the Laboratoire de Biomathématiques et d'Estimations Forestières, Faculté des Sciences Agronomiques, Université d'Abomey-Calavi in Benin.

R. Glèlè Kakaï is Professor, Laboratoire de Biomathématiques et d'Estimations Forestières, Faculté des Sciences Agronomiques, Université d'Abomey-Calavi in Benin.

Asma Hamad is with the Ministry of Production and Economic Resources, North Kordofan State, Sudan.

A. Hounkpèvi is with the Laboratoire de Biomathématiques et d'Estimations Forestières, Faculté des Sciences Agronomiques, Université d'Abomey-Calavi in Benin.

R. Idohou is Senior Lecturer, Laboratoire de Biomathématiques et d'Estimations Forestières, Faculté des Sciences Agronomiques, Université d'Abomey-Calavi in Benin.

Ahmed Idris is with the Faculty of Natural Resources and Environmental Studies, Peace University, *El Fulla*, Sudan.

Agathar Kamota is with the Department of Crop Science at Bindura University of Science Education in Zimbabwe.

Apollo Kasharu is with Chain Uganda, an NGO working with farmers.

Osman Khmis is with the Faculty of Natural Resources and Environmental Studies, Peace University in Sudan.

Dora C. Kilalo is with the Department of Plant Science and Crop Protection at the University of Nairobi in Kenya.

Friday N. M. Kubiku is with the Department of Agricultural Economics and Development, Manicaland State University of Applied Sciences.

Jimmy Lamo is Principal Research Officer at the National Crops Resources Research Institute in Uganda.

Megan Lindow is a writer focused on education, research, agriculture, food systems and sustainable development.

Ronald Mandumbu is an Associate Professor in the Department of Crop Science at Bindura University of Science Education in Zimbabwe.

Joseph W. Matofari is from Egerton University in Kenya.

Onkgolotse Moatshe-Mashiqa is with the Ministry of Agriculture in Botswana.

Ali Mohamed is with the Center for Honeybee Research, Alsalam University, West Kordofan, Sudan.

Igbal E. I. Mohamed is with the Faculty of Agricultural Science, University of Gezira, Sudan.

E. M. Mohammed is with the Department of Agricultural Extension & Rural Development, Faculty of Agriculture, University of Khartoum in Sudan.

Hana Mohammed is with the Faculty of Natural Resources and Environmental Studies, Peace University in Sudan.

Patience Mshenga is a Professor of Agribusiness Management at Egerton University in Kenya.

Johnny Mugisha is a Professor of Agricultural Economics at Makerere University in Uganda.

Basil Mugonola is an Associate Professor in the Department of Rural Development and Agribusiness at Gulu University in Uganda.

Kay Muir-Leresche is an agricultural and natural resource economist and former Chairperson of RUFORUM's Technical Advisory Committee.

Richard M. S. Mulwa is Deputy Vice-Chancellor of Administration, Planning and Development at Egerton University in Kenya.

Nancy Mungai is a Professor of Soil Science at Egerton University in Kenya.

John Mupangwa is with the Department of Animal Production, Agribusiness and Economics, School of Agriculture and Fisheries Sciences, University of Namibia.

Lubna M. Musa is with the Faculty of Agricultural Science, University of Gezira, Sudan.

Mariam Mwangi is Associate Professor, Crops, Horticulture and Soils Department of Egerton University in Kenya.

Anthony Mwangi Kibe is an Associate Professor of Agronomy at Egerton University in Kenya.

Agnes Mwang'ombe is a Professor of Plant Pathology at the University of Nairobi in Kenya.

Hikevali Ndinomuwa is with the Department of Animal Production, Agribusiness and Economics, School of Agriculture and Fisheries Sciences, University of Namibia.

Elly Kurobuza Ndyomugyenyi is Associate Professor, Department of Animal Production and Range Management at Gulu University in Uganda.

Agnes Oywaya Nkurumwa is Senior Lecturer, Agriculture Education and Extension Department of Egerton University in Kenya.

Wadzanayi Innocent Nyakudya is with the Department of Crop Science at Bindura University of Science Education in Zimbabwe.

George Nyamadzawo is a Professor in the Faculty of Agriculture, Environment and Food Systems, University of Zimbabwe.

Walter Odongo is a Lecturer in the Department of Rural Development and Agribusiness at Gulu University in Uganda.

Joshua O. Ogendo is from Egerton University in Kenya.

Daniel Ogwang is Programs Director for IIRR Uganda, an international rural development organisation.

Patrick Okori is Executive Secretary of RUFORUM and formerly served as a Principal Scientist for the International Crops Research Institute for the Semi-Arid Tropics (ICRISAT).

Duncan Ongeng is Dean of the Faculty of Agriculture and Environment at Gulu University in Uganda.

Samwel Onyango is in the Department of Food Science, Nutrition and Technology at the University of Nairobi in Kenya.

H. A. H. Osman is with the Department of Botany and Agricultural Biotechnology, Faculty of Agriculture, University of Khartoum in Sudan.

K. V. Salako is Associate Professor, Laboratoire de Biomathématiques et d'Estimations Forestières, Faculté des Sciences Agronomiques, Université d'Abomey-Calavi in Benin.

Jan Swanepoel is an Associate Professor in the Department of Sustainable Food Systems and Development at the University of the Free State in South Africa.

Bamphiti Tiroesele is with Botswana University of Agriculture and Natural Resources.

Cecil Togarepi is with the Department of Animal Production, Agribusiness and Economics, School of Agriculture and Fisheries Sciences, University of Namibia.

Peter Wasswa is from the Department of Agricultural Production at Makerere University in Uganda.

INTRODUCTION

Anthony Egeru, Patrick Okori and Megan Lindow

Introduction

Universities are well positioned to play a catalytic role in rural development through agriculture and food systems in the global south. In Africa, many universities have struggled to fulfil this potential and remain stuck in 'ivory tower' mode. This book proposes a method to enable African and global south universities to break free of the ivory tower and reposition themselves through structured engagements with communities and other societal actors that are both practical and transformative. The essential elements of this approach are articulated below:

- **Community action research platforms (CARPs)** are multidisciplinary, multi-stakeholder innovation platforms intended to catalyse on-the-ground rural transformation through participatory research and capacity building in targeted smallholder agricultural value chains to improve livelihoods providing opportunities for locally relevant research and experiential learning.
- **CARPs are innovation platforms**, collaborative arrangements involving a variety of different community, academic and non-academic stakeholders with the purpose of producing innovations to increase agricultural productivity and strengthen commodity value chains.
- **CARPs are interdisciplinary**, meaning that they integrate a variety of academic disciplines to ensure that various land-based, production, post-harvest, marketing, processing, nutrition, and organisational issues are addressed comprehensively rather than in isolation.
- **CARPs are multi-stakeholder**, meaning that each platform brings together different actors from local and national governments, Technical Vocational Educational Training institutions (TVETs), extension services, non-governmental

DOI: 10.4324/9781003387497-1

organisations (NGOs) and the private sector with the universities and communities engaged in the value chain in the research area.
- **Value chain** is a term that encapsulates all the activities involved in producing a particular agricultural commodity from soils to nourishment, including crop and animal husbandry, planting, harvesting, processing, value-addition, marketing and consumption.

Universities have the potential to serve as fulcrums of development in their societies if they align their research towards a transdisciplinary and collaborative vision that brings together students, communities, governments, tertiary training colleges, civil society and the private sector in the development of agricultural value chains. Historically, the colonial era design of many traditional African universities, combined with serious underfunding and weak capacity building, has made it difficult for these institutions to serve the development needs of their societies. As a result, African universities now face a crisis of relevance, particularly as they struggle to address the needs of the continent's rapidly growing young populations (Kraybill, Lynam and Adipala, 2021). Addressing this context, this book explores the evolution and potential of a pathway to relevance for African universities which positions students, researchers and institutions alike as change agents, facilitators of development and 'scaling laboratories' supporting rural communities in taking up useful innovations to create new industries and transform their livelihoods.

The CARP methodology developed within the Regional Universities Forum for Capacity Building in Agriculture (RUFORUM) which is a pan-African network of more than 160 universities focused on human capacity building and institutional strengthening in agriculture and food systems. Since its inception in 2004, RUFORUM has championed the role of African higher education to policymakers and development partners while simultaneously working to transform the sector with an emphasis on student-centred experiential learning, community engagement and participatory action research. Over time, these efforts have focused increasingly on inculcating entrepreneurship in students, institutions and rural communities in order to stimulate much-needed agro-industrial growth (Woldimichael et al., 2017). This book presents a rich array of stories and case studies sharing lessons of how the CARPs developed and how they have catalysed a multitude of different impacts and spinoffs across diverse settings and contexts.

Highlighting the experiences of the CARPs, this book seeks to demonstrate the powerful effect universities can have when they focus their efforts in training, research and innovation on developing targeted rural value chains with communities and wider stakeholders. Relationship building and the facilitation of partnerships among communities, researchers, government and the private sector is key to the approach. This strong focus on multi-sectoral relationships both strengthens the capacities of communities to ultimately drive their own agendas and connects them to other value chain actors whose knowledge, perspective and resources can help them grow and sustain their efforts over time. As this book illustrates, the CARP approach positions universities to contribute meaningfully towards meeting Africa's (and the world's) most pressing development challenges at the community level, including food and nutrition security, climate change adaptation, youth employment, rural livelihoods, resilience to drought and other areas of vulnerability. Positioning themselves as active and engaged facilitators of development, universities have an essential role to play in responding to virtually all the Sustainable Development Goals, particularly through their contributions towards promoting skills for adaptation to a rapidly changing world and capacities for job creation among youth – a full two-thirds of whom are either unemployed or in vulnerable employment (Nwokolo, 2022).

This book chronicles the extensive body of knowledge that has been developed through RUFORUM's long experience of nurturing student-centred, participatory and community-engaged approaches in African higher education. RUFORUM's own history and identity as an African-led university network is deeply intertwined with these ongoing efforts. RUFORUM developed out of a Rockefeller Foundation (1992–2003) initiative to support several universities to pilot small competitive graduate research grants (GRGs) for field-based experiential and participatory research with communities. Recognising the deep and transformative impacts of these limited interventions on their institutions and communities, RUFORUM's founders established the network in 2004. In 2010 RUFORUM added the CARPs to their competitive grants portfolio. The expanded scope and design included a new focus on value chains and multi-stakeholder innovation platforms. As the coordinating agency for the CARPs, the RUFORUM Secretariat played a key role in mentoring, supporting, training and backstopping researchers and universities as they tested and experimented with this novel approach. As described in Chapter 1, five early CARP projects were successfully piloted (2010–2016).

In 2016, RUFORUM launched a new iteration of the CARPs under its flagship programme, Transforming African Agricultural Universities to meaningfully contribute to Africa's growth and development (TAGDev), with support from the Mastercard Foundation. This initiative contained two parts. Firstly, it focused on two institutions: Egerton University in Kenya and Gulu University in Uganda, to develop new models of higher education that are more inclusive, build young peoples' skills as entrepreneurs and involve universities in multi-stakeholder collaborations and mutual learning across innovation systems (Lindow, 2022). The experiences of both institutions are highlighted in Chapter 6 of this book. Related to this, the TAGDev programme also provided for a new round of 16 CARP projects at different RUFORUM member universities in Benin, Botswana, Ghana, Kenya, Namibia, South Africa, Sudan, Uganda and Zimbabwe. These new CARP projects built further on the learnings already gleaned from RUFORUM's participatory, field-based approaches to training and research. They also brought a more explicit focus on stimulating entrepreneurship, strengthening relationships with the

TVETs and helping transform the lives of marginalised youth. While three of the early CARPs[1] are covered briefly in Chapter 1, the subsequent round of CARPs, operating under the TAGDev (2017–2024), are the main focus of this book.

The wider aim of the TAGDev investment by the Mastercard Foundation was to broaden access to education for the excluded and to promote new models of higher education that focused on key issues of institutional relevance and entrepreneurial education – transforming the universities into fulcrums for development. This formed part of the programme's overarching aim of strengthening the capacities of young people to become entrepreneurs and job creators, and to drive the growth of small-scale agro-industrialisation, recognised as a key sector with the potential for lifting millions of rural people from poverty (Juma, 2012; Dorosh and Thurlow, 2018; Woldimichael et al., 2017). It was envisioned that students and researchers working in transdisciplinary and multi-stakeholder collaborations would gain critical skills in facilitation, leadership, creative problem-solving and entrepreneurship. These are lacking in traditional academic settings yet are critical for successful graduates and rural enterprise development. These multi-agency community action research platforms build relationships that evolve over time forming partnerships and building connectivity among different value chain stakeholders to support the commercialisation of smallholder farmers and strengthen the relevance of universities to their society. Through such approaches, this book will argue, universities are well placed to connect global research with the diversity of realities on the ground, testing relevant innovations in local settings and embedding useful outcomes in sustainable platforms of stakeholders. As shown in this book, investing in postgraduate education that connects all these different aspects has the potential to significantly improve farmer livelihoods and national productivity while building human capital.

This book covers the major learnings and experiences of CARPs in two parts. Part I (Chapters 1 to 6) provides a comprehensive yet general overview of the key themes and processes of the CARPs in an accessible story style. Part I draws lessons to highlight to policymakers, development and implementing agencies, and provides evidence of the results that can be achieved when universities are supported to produce relevant graduates, pro-actively engage communities and facilitate the transformation of their own institutions and their societies. The Part I chapters are as follows:

- **Chapter 1 Designing for relevance** describes how the CARP approach evolved and shares key lessons of early platforms.
- **Chapter 2 Building value chain innovation platform engagements with communities** outlines the processes of building participatory research engagements with farmer communities.
- **Chapter 3 Universities fit for purpose** highlights how the CARPs have influenced a wider appreciation and uptake of student-centred learning and community engagement within universities.

- **Chapter 4 Learning in motion** looks at how experiential learning in multi-stakeholder platforms has driven iterative cycles of action and transformation in the CARPs that continue to build.
- **Chapter 5 Change agents, spinoffs, wider impacts** illustrates some of the emergent and unplanned impacts the CARPs have had for students and in communities and institutions alike.
- **Chapter 6 Facilitating a student and community-centred, experiential approach to research and innovation** captures the experiences of Egerton and Gulu Universities where the CARPs formed part of wider institutional change processes under TAGDev.

Part II of this book is authored by the researchers themselves and explores the impact of the individual CARPs that have been implemented 2017–2024. Part I draws upon and refers to numerous stories and examples from these projects that are outlined in greater technical detail in Part II.

The Part II chapters are as follows:

- **Chapter 7 Enhancing agribusiness rice clusters and market linkages for food security and incomes in northern Uganda** (Gulu University, Uganda; 'Uganda rice CARP' in Part I).
- **Chapter 8 Enhancing safflower production and product development for food security and improving incomes of small-scale farmers in Botswana** (Botswana University of Agriculture and Natural Resources; 'Botswana safflower CARP' in Part I).
- **Chapter 9 Sustainable commercial pineapple value chain system for increased yield and income, and improved livelihood of smallholder farmers in Central Region, Ghana** (University of Cape Coast, Ghana; 'Ghana pineapple CARP' in Part I).
- **Chapter 10 Empowerment and poverty reduction in rural coastal Kenya through the cassava value chain** (University of Nairobi, Kenya; 'Nairobi cassava CARP' in Part I).
- **Chapter 11 Unlocking opportunities in the cassava CARP value chain in Nakuru County, Kenya** (Egerton University, Kenya; 'Egerton cassava CARP' in Part I).
- **Chapter 12 Seed potato production in Nakuru, Kenya: outcomes and implications of an active multi-stakeholder platform** (Egerton University, Kenya; 'Egerton seed potato CARP' in Part I).
- **Chapter 13 Inclusion of farmer communities in value chain development: a case of potato in southwestern Uganda** (Makerere University, Uganda; 'Uganda potato CARP' in Part I).
- **Chapter 14 Enhancing pig production and marketing for smallholder farmers in northern Uganda** (Gulu University, Uganda; 'Uganda pig CARP' in Part I).

- **Chapter 15 Building competitiveness for communal farmers through developing the wool value chain in the free state province of South Africa** (University of the Free State, South Africa; 'South Africa wool CARP' in Part I).
- **Chapter 16 Climate-smart agriculture: improving dryland crop yields and value addition through university-community partnership in Zimbabwe** (Bindura University of Science Education, Zimbabwe; 'Zimbabwe CARP' in Part I).
- **Chapter 17 Enhancing community adaptation through climate resilient agriculture: encroacher bushes value chains initiative in Otjozondjupa Region, Namibia** (University of Namibia; 'Namibia bush CARP' in Part I).
- **Chapter 18 The empowerment of women and youth: enhancing climate-smart productivity of vegetables and key livestock value chains, Sudan** (University of Gezira, Sudan; 'Sudan women CARP' in Part I).
- **Chapter 19 Strengthening the beekeeping value chain in Western Kordofan State, Sudan** (Peace University, Sudan; 'Sudan beekeeping CARP' in Part I).
- **Chapter 20 Improvement of indigenous coping strategies in famine-stricken Darfur, Sudan** (University of Khartoum, Sudan; 'Sudan famine foods CARP' in Part I).
- **Chapter 21 Scaling up African baobab food products valuation through enhancement of their safety and value chains for food and nutritional security in Benin, West Africa** (University of Abomey-Calavi, Benin; 'Benin baobab CARP' in Part I).

BOX 0.1 RUFORUM AND THE TAGDEV PROGRAMME THAT IMPLEMENTS THE COMMUNITY ACTION RESEARCH PLATFORMS

RUFORUM – the Regional Universities Forum for Capacity Building in Agriculture – is a network of over 160 universities in Africa. Its mission is to strengthen the capacities of universities to foster innovations responsive to the demands of smallholder farmers and value chains through the training of high-quality, relevant graduates, the output of impact-oriented research, and the maintenance of collaborative working relations among researchers, farmers, market actors, national agricultural research and advocacy institutions, and governments. The network, owned by the member universities, was established in 2004 by ten Vice-Chancellors building on a project that linked African universities to each other and supported faculty and students to improve their skills and engage in participatory research. RUFORUM is registered as an NGO in Uganda where the Secretariat is based. Every year members gather[2] at different venues across Africa sharing ideas, showcasing innovations and advocating for the transformation of higher education and agriculture.

Although their faculty and students compete for grants within their universities and across the network, the university members actively support each other and collaborate closely in research, training and advocacy.

Transforming agriculture in Africa requires innovative scientific research, student-centred, experiential learning and universities embedded within society using new technology and collaborative partnerships. Tertiary education needs to be more connected to the new challenges facing rural communities and build the skills and networks of young people to drive adaptation of innovations and establish businesses that advance transformation.

One of the RUFORUM Flagship Programmes is 'Transforming agricultural universities to meaningfully contribute to Africa's Growth and Development' (TAGDev). It supports the RUFORUM commitment to inclusive and sustainable education and agriculture. It broadens access and enhances the relevance and quality of graduates and innovations in rural communities. The Community Action Research Platforms (CARPs) are one of the initiatives under this programme. They strengthen the facilitation role of universities using their neutral convening power to bring together stakeholders across a value chain to collaborate with the university and smallholder farmers to enhance rural agro-industrialisation.

The TAGDev programme is an initiative co-created with, and financially supported by, the Mastercard Foundation. It is implemented and shepherded by the RUFORUM Secretariat with the support of experts from member universities and the RUFORUM Technical Committee. This book presents some of the results and lessons from these CARPs. Part 1 provides an overview and uses some of the stories of different stakeholders. Part II outlines the processes and impacts of each CARP as told by their platform initiators.

Further information may be accessed here:

- www.ruforum.org
- www.ruforum.org/MCF/about-tagdev-0
- https://ruforum.org/impact/

Notes

1 Two other CARPs were initiated. One in 2014, Mekelle Ethiopia focused on wheat but the sustainability of its platform was stymied by ongoing conflict in the northern Tigray region of Ethiopia. The dairy CARP in Tanzania initiated in 2015 also had some early successes.

2 https://ruforum.org/ruforum-conferences.

Bibliography

Dorosh, P. and Thurlow, J. 2018. 'Beyond Agriculture Versus Non-agriculture: Decomposing Sectoral Growth–Poverty Linkages in Five African Countries'. *World Development*, 109: 440–445.

Juma, C. 2012. 'Building New Agricultural Universities in Africa'. Faculty research working paper series. John F. Kennedy School of Government, 12(26). https://dash.harvard.edu/handle/1/9359206.

Kibwika, P. 2006. *Learning to Make Change: Developing Innovation Competence for Recreating the African University of the 21st Century.* Wageningen Academic Publishers.

Kraybill, D., Lynam, J., and Adipala, E. 2021. *Transforming Tertiary Agricultural Education in Africa.* CABI, Oxfordshire.

Lindow, M. 2022. 'The RUFORUM Journey'. RUFORUM. PDF. http://repository.ruforum.org/system/tdf/The%20Ruforum%20Journey.pdf?file=1&type=node&id=40580&force=.

Nwokolo, M. N. 23 February 2022. 'Youth Unemployment is Africa's Biggest Crisis'. *Brenthurst Foundation.* https://www.thebrenthurstfoundation.org/news/youth-unemployment-is-africas-biggest-crisis-and-puts-the-continent-at-risk-of-becoming-home-to-centres-of-chaos/.

RUFORUM. n.d. 'About TAGDev'. Website. https://www.ruforum.org/MCF/about-tagdev-0.

Woldimichael, A., Salami, A., Mukasa, A, Simpasa, A., and Shimeles, A. 2017. 'Transforming Africa's Agriculture through Agro-industrialization'. *Africa Economic Brief*, 8(7). African Development Bank. https://www.afdb.org/fileadmin/uploads/afdb/Documents/Publications/AEB_Volume_8_Issue_7_Transforming_Africa_s_Agriculture_through_Agro-Industrialization_B.pdf.

PART I

Lessons and voices from the CARPs

1

DESIGNING FOR RELEVANCE

A new orientation for research

Anthony Egeru, Megan Lindow and Patrick Okori

Introduction

This chapter explores the evolution of an approach to building collaboration between African universities, smallholder farmer communities and a wide variety of other institutional, NGO and private sector partners to address deep rural development challenges and overhaul student training. Community action research platforms (CARPs) are innovation platforms designed to reposition African universities to catalyse sustainable and inclusive development in the agrifood sector.

Small-scale farmers are the backbone of African societies, yet over many decades institutions have ignored their needs, leaving them disproportionately impoverished and food insecure (Lynam et al., 2016; Fanzo et al., 2020; Mazungunye, 2020; Nkhono-Mvula et al., 2023). The CARP approach seeks to demonstrate the potential of collaboration, driven by the untapped brain power of the universities, to begin to reverse decades of underdevelopment and catalyse small-scale agro-industrialisation and development in rural areas. Essentially, this approach involves repositioning universities for greater relevance to, and impact within, their societies.

The CARP approach proposes a radical departure from the roles for which most traditional African universities were designed. Established from the early 1900s onwards, these institutions were often envisioned as ivory towers, created largely to serve colonial administrations and, later on, the bureaucracies of independent African nation states by providing suitable employees for civil administration. The mode of learning was to prepare functionaries for white-collar office jobs, not to fledge entrepreneurs to innovate, disrupt and drive change. So long as the civil services absorbed most university graduates, there was little incentive for the universities to change and little room for them to manoeuvre either (Kibwika, 2006).

DOI: 10.4324/9781003387497-3

As institutions faced worsening problems of underfunding and neglect, there was a need to think differently. Universities came under immense pressure from the public to demonstrate their relevance and governments started asking difficult questions associated with value for money and the impact of universities in development. African universities largely avoided such conversations over the last century and into the first quarter of the 21st century – even as their ranks of jobless graduates multiplied. Many universities today are still waking up to the public's demand that they prove their relevance. Sooner rather than later, they will need to answer.

By the turn of the 21st century, the magnitude and complexity of Africa's environmental, economic, population and development challenges had rapidly outstripped the universities' capacities to engage with them. The inherited ivory-tower design limited the scope of universities for community engagement and knowledge co-creation, and the universities found themselves increasingly on the margins of their rapidly changing societies. Structural adjustment and declining budgets together with massive expansion led to overcrowding and heavy workloads for faculty, reducing both time and financial resources for intensive engagement with rural communities. Economic growth and agricultural productivity stagnated for several decades (Juma, 2012; Myeki et al., 2022). The ranks of jobless university graduates proliferated, as studies and employer surveys revealed graduates' lack of the critical skills demanded by the labour market (Kibwika, 2006). Today, as nearly two-thirds of African youth find themselves either unemployed or in vulnerable employment, universities have done little to inculcate entrepreneurship. Yet as agricultural growth gains recognition for its unmatched potential to lift people out of poverty and the global Sustainable Development Goals focus efforts on ending poverty and hunger and creating dignified livelihoods within environmental limits, a new pathway to relevance for African universities has opened (Juma, 2012; Adipala et al., 2021; Adesina, 2018).

Developing an iterative, holistic and collaborative research approach

The CARPs are a product of an evolution in thinking and practice within RUFORUM, a pan-African network of universities focused on capacity building in agriculture and food systems. Since its inception in 2004, RUFORUM has employed small grants to provide opportunities for Participatory Action Research to train master's students and to support faculty to work with rural communities and engage with other stakeholders (Mapfomo et al., 2013). The PAR approach was adopted as part of RUFORUM's broader efforts to overhaul postgraduate training and empower young scientists as proactive change-makers focused on delivery and impact. With participatory change-making at its core, PAR was seen as an effective approach to engage researchers in understanding and addressing the complex and multi-faceted challenges faced by smallholder farmer communities (Kibwika, 2006; Kindon et al., 2007; Chevalier, 2019). Graduate research grants focused on field-based research and training for MSc students; they also gave university staff experience in managing research grants and strengthening student supervision and mentoring. All while developing and delivering knowledge and innovations that respond to smallholder farmer needs. For many researchers, these grants offered a unique opportunity to step out of the confines of the lab and engage with communities. PAR has been used in RUFORUM's competitive grants processes since its inception (see Box 1.1).

RUFORUM's emphasis on PAR through their Graduate Research Grants and Field Attachment programmes was in marked contrast to what was happening in most universities before 2000. Their postgraduate research focused on university or research station trials and laboratories, and seldom involved engagement between students and farmers. However, more interdisciplinary and farmer-oriented research approaches began to take root for smallholder farms in the 1980s in Zimbabwe with the introduction of a Farming Systems Research approach that was also characterised by action research (Rukuni et al., 1998; Chambers and Jiggins, 1987). Towards the end of the 20th century, more community-engaged systems with some focus on smallholder farming by research agencies using participatory action research methods (PAR) was emerging.

From its inception, RUFORUM actively supported this approach. The CARPs, emerging from the lessons of implementing smaller Graduate Research Grants that used PAR (as described above), both took inspiration from and expanded these earlier approaches to a new level. Essentially, the CARP approach harnesses and extends the PAR approach in a unique way that establishes platforms of engagement, linking farmers to wider systems and building relationships that are mutually sustaining. The CARP approach centres around post-graduate students working with their supervisors in research teams that engage directly with communities to determine research priorities. It provides opportunities for participatory action research and experiential learning, training students with relevant, development-oriented skills in the process. But it is more than conventional action research because it relies on building an innovation platform that engages with stakeholders from across commodity value chains. It links the farmers, the universities and technical and vocational colleges (TVETs) with local and national government agencies, civil society organisations, international research agencies and the private sector. It builds relationships and strengthens networks and impact.

The book explores the experiences and lessons drawn from 21 different projects initiated by RUFORUM member universities in Benin, Botswana, Ethiopia, Ghana, Kenya, Malawi, Namibia, South Africa, Sudan, Tanzania, Uganda and Zimbabwe. It highlights a critical yet under-utilised role for African universities in stimulating rural economies. It argues for recognition of universities' important roles as collaborators and catalysts for multi-sector solutions (Juma, 2012). These are solutions that increase productivity, address climate change, develop livelihoods and resilience in rural communities and promote farmers' access to markets, innovation and trade, while safeguarding biodiversity and enhancing food and nutrition security.

BOX 1.1 THE POPULAR KNOWLEDGE WOMEN'S INITIATIVE (P'KWI)

An important seed of the CARP approach was born in the mid-1990s, out of a Rockefeller Foundation-supported grants programme focused on postgraduate training in agriculture at several African universities. The programme, called the Forum for Agricultural Resource Husbandry (FORUM), was the predecessor to RUFORUM. A small research team awarded under FORUM and led by plant pathologist Professor Ekwamu Adipala from Makerere University in Uganda was working to develop new varieties of cowpeas, an important dryland food security crop. They wanted to learn the particular qualities of cowpeas that farmers would find most useful and desirable so that they could breed selectively for those attributes. The researchers connected with the Uganda Ministry of Agriculture and with research stations in three cowpea-growing districts of eastern Uganda who introduced them to several different farmers' groups. The area was emerging from civil conflict. Farming had been severely disrupted and communities recovering from the trauma faced deep food insecurity.

A farmer named Norah Ebukalin, along with her husband, the Reverend Sam Ebukalin, had begun to organise a small group of local women to revive farming and rebuild their lives. They named themselves the Popular Knowledge Women's Initiative (P'KWI). When researchers connected with the women's group and proposed collaboration, both parties were eager to work together but had to find their way forward.

The researchers were able to provide the community with much-needed cowpea and groundnut seeds, cassava stems and basic farming tools. Master's students working on different aspects of plant breeding, crop disease and pest management and socio-economic issues planted and tested different varieties with farmers in their fields. This collaborative way of working produced valuable mutual learning and insight for the farmers and the researchers. As Adipala and Lindow (2019) wrote:

> Working side by side with farmers, we could appreciate the deep knowledge they held. We were not the experts here in the field: the farmers were. They were the ones who knew how to space their plantings of groundnuts in such a way that the aphids would not devour them. They were the ones who knew how to protect their crops from pests without using chemicals. They were the ones who really understood which cowpea varieties – and later on, which sorghum varieties – would grow best in their home soil. It was very humbling to see how much knowledge these farmers possessed, once we researchers developed the eyes to see what they were capable of and how we could support these efforts.

Nearly 30 years later, P'KWI is a successful farmer group of more than 2,500 households which continues to embrace research, innovation and learning to further its development. Today, the group produces organic sunflower oil in a collectively owned modern factory, among other initiatives. Recently Norah Ebukalin (2022) reflected:

> I am very proud of what we have done. Looking back, we made a marriage between indigenous knowledge and scientific knowledge, meaning that some of the things we were doing that were not evidence based could be researched. The knowledge was on the ground and could be used in the right way, unlike before when knowledge remained in the archives and demonstration plots of the universities. We mixed these approaches and were able to find a solution to the pests in the garden without using pesticides. There are some things the classroom cannot teach you, but if you do research in the communities, you can find the solutions working together.

Community action research platform design to address farmers' challenges

CARPs have been applied in very different value chains – involving diverse types of platforms and stakeholders. The core aim throughout is experiential learning for the students in researching and working collaboratively with the community being served. The Principal Investigator, researchers and mentors support the students and work with other stakeholders to ensure the potential for uptake of new ideas in rural communities. They strengthen the individuals and institutions involved.

The CARP approach extends this research into a use concept. It is part of a wider, gradual paradigm shift towards experiential learning and knowledge co-creation in African higher education and offers a rich source of context-specific learning and experience in working with the communities from the onset to establish the kinds of interventions required to address rural agricultural value chain development. The approach stimulates change in the universities and in rural areas by opening relationships and pathways for uptake. It positions students, researchers and universities themselves as facilitators of change and fulcrums of development in their societies. These aspects are explored in more depth in Chapters 3, 4 and 5.

Aims of the Community Action Research approach

While African governments have provided a policy environment to support the proliferation of small-scale agro-industries, practical work on the ground with communities in this direction has been limited (Kovach, 2020). However, there is a renewed commitment and interest in accelerating agro-industrialisation in

Africa because it is considered as the primary route for broad-based economic development and the first stepping stone in the industrialisation process (Ouma and Whitfield, 2012). Various policy frameworks including the African Union's Agenda 2063 framework, as well as national and local policy instruments of Kenya, Malawi, Zimbabwe and other countries, all highlight the critical role of agro-industrialisation to support national and continent-wide development agendas. However, universities and other institutions have struggled to coordinate their efforts on the ground and produce the mechanisms to catalyse and support such growth in communities. The same underinvestment and structural adjustments that weakened higher education institutions across Africa during the 1980s and 90s also took their toll on national agricultural research systems, extension services and TVETs. As a result, small-scale farmers in many communities have often been left with piecemeal and sporadic support systems and a lack of coordinated effort to help them move forward.

The CARP approach sought to overcome these structural obstacles through practical collaborative engagement. The approach follows an iterative process that begins in a dialogue with communities to establish research problems they can address together. In the process, discussions between communities and researchers begin to shape an understanding of the challenges and opportunities of the landscape and, from this, a research agenda begins to take form.

The CARP specifically seeks to:

1. Support collaborative action research led by universities with the participation of graduate students that addresses critical agricultural production, processing and marketing constraints facing smallholder farmers in Africa addressing inclusive rural livelihoods, food and nutrition security and climate change;
2. Provide opportunity for the creation of multi-stakeholder 'platforms' that support value chain development for the benefit of smallholder farmers and to enhance the impact of university activities;
3. Provide participatory action research platforms and opportunities for greater student engagement and experiential learning in supporting smallholder farm development through agribusiness and ensure that lessons are used to facilitate university reflection and learning;
4. Support strengthening of the educational value chain and in particular technical and vocational (TVET) institutions through engagement and joint action between them and universities to enhance uptake of research and innovations on smallholder farms.

Each CARP normally includes scholarships, research funds and training for at least one doctoral and four master's students with undergraduates as research assistants and TVET college students also engaged. The students are linked with small farmers, agri-entrepreneurs and community organisations to carry out research with them in cooperation with other agencies. They identify key constraints along the value chain and engage with all stakeholders: local and national government, research, extension and development agencies, civil society and the private sector. The approach involves the Principal Investigator of each CARP to facilitate and coordinate both the academic and the outreach activities of this multi-disciplinary, multi-agency platform. Each CARP implements its participatory action research (PAR) projects through the establishment of an outreach or innovation platform. This links the universities with rural stakeholders through close collaboration with a wide range of commercial, research, development, government and civil society agencies servicing the communities.

The CARPs establish a multi-stakeholder platform of engagement creating dynamic networks that continue to operate after the project is over. Furthermore, the CARPs encourage the students and farmers to take up the new technologies and create enterprises that help to scale out the innovations. They require the researchers to mainstream gender and the environment into their projects. A CARP also deliberately addresses links between universities and technical colleges. The Principal Investigator leading a CARP needs to focus on facilitating broad stakeholder engagement. The approach establishes very specific milestones that all partners are required to adhere to which has assisted with streamlining some of the academic processes, strengthening market links and impacting policies.

The CARPs assist universities to improve their ability to identify research priorities that reflect farmer and community priorities. Research is based on local issues and challenges so that CARPs enhance the universities' ability to learn and adapt existing and ongoing research and innovations for greater uptake and to adapt curricula to produce more relevant graduates. The approach stimulates change in the universities and in rural areas as the engagements among stakeholders continue to generate new pathways for further innovation and uptake. Driven by such continuous iteration, the intention of the design is that the innovation platforms provide a context for the universities to serve as fulcrums of development. The stories in this book show that the platforms have helped universities to support both government and civil society research and development agencies, have enhanced private sector engagement, have developed small businesses run for and by students and communities and have improved the competitiveness of African products. Universities are adapting new learning approaches through the CARPs. This book helps to scale out the lessons learned from the CARPs into other university research, learning approaches and curricula.

How have the CARPs evolved?

As detailed above, the CARPs were an initiative of the RUFORUM network to scale up the impacts of its Graduate Research Grants (GRGs). It was observed that while the small grants programmes were effective as mechanisms for training, they were too limited in their scale and duration to have a lasting impact on communities. They were too small to include the wide range of stakeholders needed to effectively address the full value chain. Universities and communities continued to exist in their separate spheres. Some of these small grants did trigger follow-up research that eventually made significant advances such as soya beans in Zimbabwe and Uganda and the cowpeas (see Box 1.1). Their success demonstrated the potential of participatory action research with communities to catalyse deeper and more lasting impacts if the mechanisms of collaboration could be strengthened.

In 2010, RUFORUM managers and representatives from member universities and governance organs brought forward a proposal to expand the small graduate research grants into more comprehensive research platforms. The Bill and Melinda Gates Foundation incorporated funding for three Community Action Research Projects into their funding grant to RUFORUM and a Competitive Call went out to the network. The proposals were to include PhD and MSc students, using undergraduate interns to assist during their vacations. They were to be multi-disciplinary and multi-agency and had to address issues identified with local rural communities. They needed to be innovative and establish a collaborative approach to encourage the uptake of technologies and result in increased livelihoods that were sustainable and that included women researchers and farmers. From 2010 to 2016, five five-year CARPs with grants of US$300,000 each were established in Uganda, Kenya, Malawi, Ethiopia and Tanzania.

One such grant was awarded to Makerere University in Uganda to focus on preventing diseases and developing opportunities in the cassava value chain. The project engaged several farmer groups and built further on earlier farmer relationships established under several previous GRGs. The project focused on making quality, disease-resistant cassava varieties, including seeds and stems, available to farmers and distributing them at an affordable cost. The project also focused on value addition. As Principal Investigator Professor Settumba Mukasa recalls, farmers gave him valuable insights about the multiplicity of challenges they faced, and how to navigate expectations and build relationships in the community. As a virologist previously confined to working in the laboratory, he had a steep learning curve ahead of him.

One of the communities involved in the project suffered from drought which produced a major food insecurity crisis. With little greenery left except for mango trees and the cassava stems planted in the project's mother garden, farmers' cattle quickly found their way to these green cassava stems and demolished them. 'These were not situations you encounter in the lab', Mukasa recalls. 'We had been taught that cassava is drought tolerant. But I had never thought about how drought plus cows plus the diseases we were fighting would affect our local cassava source'. Constant engagement was needed to understand the nuances of what the communities really wanted and needed from the project as they together explored different value chain avenues, including the production of bio-ethanol from under-utilised dried cassava chips. Mukasa (2022) reflected:

> The community engagement allows you to understand all these drivers you otherwise would not understand. In terms of the bio-ethanol production, we learned

> that those who produce the alcohol make a lot of money, and this drives the demand for cassava flour. On the one hand, it is alcohol production and many people disapprove of it. Yet it's also an important part of the local economy and has many advantages, such as being able to store it indefinitely.

Ultimately, farmers who participated in the CARP saw many benefits, including better networks and increased access to planting materials, inputs, markets, processing facilities and seeds. For the researchers, continuous engagement with the community was key and provided rich learnings. As Mukasa (2022) recalls: 'Triangulation was key to the research. You could ask farmers what they thought were the biggest problems, but there were always different perspectives among the men and women, or the young versus the old. It was a big puzzle'.

Another CARP grant was awarded to the University of Eldoret in western Kenya, where researchers worked with farmer groups towards the development of strong farmer cooperatives to address common challenges. It started with the challenge that simple and affordable technologies produced by the university to improve the quality of local farmers' soils and yields were not being adopted by the communities for which they were intended. Researchers believed the failure of uptake was partly because the university did not have strong relationships with farmer groups and farmer groups themselves had weak capacities. Farmer cooperatives for grain, coffee and cotton growers had once been widespread across Kenya but had succumbed to the collapse of global commodity prices during the 1980s (Lindow, 2014). Government extension services had similarly dwindled. To bridge these institutional gaps, the CARP sought to engage communities in collaborative learning processes, as opposed to a traditional top-down extension model, exploring how farmer groups could organise and strengthen themselves to address production, management and marketing holistically along entire value chains.

Yet part of the initial problem of uptake also lay with the traditional structures and modes which positioned the researcher as the expert. This separated their expertise both from the wider knowledge systems of the university and community and from the wider social contexts in which the innovations might be useful. The CARP mobilised a transdisciplinary team of researchers and students to support farmer groups with combining and optimising different technologies and interventions to improve their yields, control the vociferous striga weed that spread across their land, develop value addition and reach new markets with better opportunities (Lindow, 2014). Addressing these combined challenges in a holistic collaboration meant that farmers could move faster to overcome their challenges and work together in a broader set of relationships to develop new opportunities.

Three major lessons were learned through the University of Eldoret CARP:

1. Farmers' willingness to join an association depends on how they perceive the benefits of membership, transparent systems and ultimately on having solid management structures in place.

2. Farmer-to-farmer training and mentorship is needed to enhance the uptake of certain technologies.
3. Serious engagement is needed among all the people and institutions involved in the collaboration.

(Lindow, 2014)

Taken together, these lessons highlighted the importance of participatory action research using a multi-stakeholder innovation platform for generating deep knowledge from different perspectives to inform interventions. The CARPs brought together students and researchers working in such diverse disciplines as agri-enterprise development, agricultural economics, marketing, disease management, soil science, food science, technical areas such as tissue culture, the socio-economics of production and gender studies.

Addressing their structural and governance challenges with CARP support, farmer organisations were able to strengthen themselves. One intervention introduced through the CARP was the Mbili system which introduced soya to traditional maize production cycles for improved nitrogen fixation in the soil. The method gave farmers a soya crop which they had not had before. Through organising as a co-operative, the project also helped farmers to develop and market value-added soya products such as soap, cake and flour. Working as a group also gave them bargaining power, resulting in bulk access to inputs and better prices from buyers (Lindow, 2014). For the university, the collaborations strengthened capacities for community engagement, transdisciplinary research and provided students and graduates with sought-after skills. Working with farmer groups, researchers and students learned to work in multidisciplinary teams, manage group dynamics and conflicts, improve their gender sensitivity and build relationships with local government, NGOs and private companies including local fertiliser manufacturers. They gained valuable facilitation, communication and problem-solving skills. Taking up learnings from the CARP, the university restructured its student practicals and placed students in attachments with local farmers, with the result that 40 percent of students are now directly employed by local industries upon graduation. As will be discussed further in Chapter 3, when the CARP programme ended, the university established an outreach centre in order to institutionalise the relationships it had built with communities and farmer groups to ensure that mutual learning continues.

A third CARP was granted to the Lilongwe University of Agriculture and Natural Resources (LUANAR) in Malawi to develop a fishponds aquaculture value chain working with small-scale farmers. As the vast Lake Malawi suffered a collapse of its fish stocks, university researchers saw an opportunity to develop new interventions to support a viable aquaculture sector and secure greater access to a key protein source for Malawi's population. Researchers had bred a faster-growing species of tilapia and introduced new techniques in fish feed and the management of aquaculture ponds. Working with farmer groups to promote the uptake of these technologies, the CARP was key to establishing the credibility of the university, upgraded from a training college under the University of Malawi to an independent university in 2012. Through the CARP, farmers gained food security and were able to improve the quality of life of their families. Meanwhile, the university attracted international students and forged relationships with policymakers, helping to pave the way for a new centre of excellence and appointment as the regional policy node of NEPAD under the African Union, as will be covered in more detail in Chapter 5.

The initial CARPs demonstrated tangible effects of greater coordination and capacity strengthening across entire value chains, both in terms of stimulating the growth of local industries and boosting food security and farmer incomes at community and household levels. On the strength of these findings, seed funding was provided in 2017 under the Transforming African Agricultural Universities to meaningfully contribute to Africa's growth and development (TAGDev) programme supported by the Mastercard Foundation to establish new CARPs to be led by universities in Benin, Botswana, Ghana, Kenya, Namibia, South Africa, Sudan, Uganda and Zimbabwe.

The CARPs support full value chains of key importance to local communities and support collaborative action in community-based agro-industrialisation. They help smallholder African farmers organise themselves and overcome historic barriers that keep them locked in poverty. The approach posits that if science is to have a catalytic effect on community development, then robust participation and strong relationships among all the key stakeholders in the value chain are essential. Building on the lessons from earlier CARPs, the newer CARPs contained two new aspects to deliberately encourage enterprise development and to include the Technical and Vocational colleges. They were intended to enable the consolidation of community-level impacts. Building on the University of Eldoret's approach, the new CARPs aimed to facilitate the establishment of community-based incubation centres. The strong relationships established with local TVETs as fundamental actors in the last-mile delivery of technologies, innovations and management practices, as will be discussed further in Chapters 3 and 5.

The wider aim of the TAGDev investment by the Mastercard Foundation was to promote new models of higher agricultural education that focused on key issues of inclusivity, institutional relevance and entrepreneurial education – transforming the universities into fulcrums for development. This formed part of the programme's overarching aim of strengthening the capacities of young people to become entrepreneurs and job creators, as well as to drive the growth of small-scale agro-industrialisation, recognised as a key sector with the potential for lifting millions of rural people from poverty. It was envisioned that students and researchers working in transdisciplinary and multi-agency collaborations would gain critical skills in facilitation, leadership, creative problem-solving and entrepreneurship that are lacking in traditional academic settings yet are critical for successful rural enterprise development.

Conclusion

As will be demonstrated further in the following chapters, the CARPs' approach has opened new pathways to relevance for universities. It has enabled universities to play catalytic and collaborative roles that have directly impacted the lives and prospects of students, researchers, farmers and other community members. Working with farmers and across sectors to galvanise small-scale industrial development, the universities are showing their potential to facilitate locally impactful solutions to the interconnected challenges of rural livelihoods, youth employment, food security, climate change and biodiversity loss. The approach has strengthened communities and universities alike. It has empowered both to play stronger roles in sustainable development, as will be seen in the stories within the chapters that follow.

Bibliography

Adesina, A. 2018. 'Unlocking Africa's Agricultural Potential to Create Wealth'. Food and Agriculture Organization (FAO) of the United Nations, Rome 2018, https://www.afdb.org/en/news-and-events/unlocking-africas-agricultural-potential-to-create-wealth-18437.

Adipala, E. and Lindow, M. 2019. 'Unfinished Journey'. RUFORUM. https://repository.ruforum.org/documents/unfinished-journey-memories-road-travelled.

Adipala, E., Swanepoel, F., Mentz-Coetzee, M., Kapfudzaruwa, F., and Muir Leresche, K. 2021. 'Agriculture and Tertiary Education Within the Context of Global and African Development Goals'. In *Transforming Tertiary Agricultural Education in Africa*, ed. David Kraybill, John Lynam and Ekwamu Adipala, 3–27. CABI, Oxfordshire. https://www.cabidigitallibrary.org/doi/book/10.1079/9781789246544.0000.

Chambers, Robert and Janice Jiggins. 1987. 'Agricultural Research for Resource-Poor Farmers: Transfer-of-Technology and Farming Systems Research'. *Agric. Admin. & Extension*, 27: 35–52.

Chevalier, Jacques M. and Daniel J. Buckles. 2019. *Participatory Action Research Theory and Methods for Engaged Inquiry*. 2nd Edition. Routledge.

Ebukalin, N. 2022. Personal Interview.

Egeru, A., Nampala, P., Okalany, E., Owuor, C., Nnam, J., Apio, J., and Adipala, E. 2016. 'Taking Pro-active actions to Enhance the Quality of African Universities Graduates for the Labor Market'. *RUFORUM Working Document Series*, 14(1): 247–256.

Fanzo, J., Namukolo Covic, Achim Dobermann, Spencer Henson, Mario Herrero, Prabhu Pingali, and Steve Staal. September 2020. 'A Research Vision for Food Systems in the 2020s: Defying the Status Quo'. *Global Food Security*, 26. DOI: https://doi.org/10.1016/j.gfs.2020.100397.

Igshaan Samuels, M., Anthony Egeru, and Prisca Mugabe. 2023. 'African Rangelands and Pastoralism in a Changing Continent: Perspectives and Opportunities'. *African Journal of Range and Forage Science*, 40(1): iii–vi. DOI:
https://www.tandfonline.com/doi/abs/10.2989/10220119.2023.2182070.

Juma, C. 2012. 'Building New Agricultural Universities in Africa'. Faculty research working paper series. John F. Kennedy School of Government, Harvard University, 12(26).

Kibwika, P. 2006. Learning to Make Change: Developing Innovation Competence for Recreating the African University of the 21st Century. Wageningen Academic Publishers.

Kindon, Sarah, Rachel Pain, and Mike Kesby (eds). 2007. *Participatory Action Research Approaches and Methods: Connecting People, Participation and Place*. Routledge, New York. https://www.routledge.com/Participatory-Action-Research-Approaches-and-Methods-Connecting-People/Kindon-Pain-Kesby/p/book/9780415599764.

Kovach, H. 2020. 'External Finance for Rural Development'. Country Case Study: Indonesia.

Lindow, M. 2014. 'From Soil Microbes to Markets: Strengthening Farmers' Organisations and Value Chains in Kenya through Community Action Research'. Eldoret University CARP Case Study. RUFORUM. https://repository.ruforum.org/system/tdf/Ruforum%20Case%20Study%20-%20CARP%20Eldoret%20Story_2014_email.pdf?file=1&type=node&id=33624&force=.

Lynam J., Beintema N. M, Roseboom J., and Badiane O. 2016. *Agricultural Research in Africa: Investing in Future Harvests*. International Food Policy Research Institute, Washington, DC.

Mapfomo, Paul, Samuel Adjei-Nsiah, Florence Mtambanengwe, Regis Chikowo, and Ken Giller. January 2013. *Environmental Development*, 5: 6–22. https://www.sciencedirect.com/science/article/abs/pii/S2211464512001224.

Mazungunye, P. P. 2020. 'Industrialisation for Economic Transformation: Economy-wide Impacts of Agro-processing Development in Tanzania'. Doctoral dissertation. Stellenbosch University, Stellenbosch University.

Mukasa, R. 2022. Personal Interview.

Myeki, L. W., Bahta, Y.T., and Matthews, N. 2022. 'Exploring the Growth of Agricultural Productivity in AFRICA: A Färe-Primont Index Approach'. *Agriculture*, 12: 1236. DOI: https://doi.org/10.3390/agriculture12081236.

Nkhono-Mvula, T., Kayuni, H. M., and Mulwafu, W. O. 2023. 'Post-Colonial Agriculture Policy and Agrarian Change in Malawi, 1964–1979'. *The African Review*, 1(aop): 1–24. DOI: https://doi.org/10.1163/1821889x-bja10054.

Ouma, S. and Whitfield, L. 2012. 'The Making and Remaking of Agro-industries in Africa'. *Journal of Development Studies*, 48(3): 301–307.

Rukuni, M, Malcolm Blackie, and Carl Eicher. 1998. 'Crafting Smallholder-driven Agricultural Research Systems in Southern Africa'. *World Development*, 26(6): 1073–1087.

2

BUILDING VALUE CHAIN INNOVATION PLATFORMS WITH COMMUNITIES

Megan Lindow, Anthony Egeru and Kay Muir Leresche

Introduction

Small-scale farmers across Africa have been stuck at the base of the food system. Relying almost entirely on rainfed agriculture, small plots of land, extremely limited support from extension services and limited access to markets, their sole option has often been to produce raw commodities at a small scale. These activities support only the most basic livelihoods. Traditional methods of agricultural research were not well suited to helping small-scale farmers overcome their challenges. Traditional academic research uses an approach that focuses on conducting trials in the laboratory and on university farms to garner research publications. This approach keeps research separate from the realities and contexts of farmers' lives and limits the knowledge contributions that farmers can make. Farmers are reduced to passive recipients of the innovations and technologies developed from research, instead of participating actively in its co-creation. As a result, research produced by traditional methods often fails to meet farmers' desires and needs and therefore remains on the shelf.

Participatory action research (PAR) is particularly suited to working with small-scale farmers (Mponela et al., 2023) because it is collaborative by design, engages with real-life situations and is specifically geared towards making change (Kibwika, 2006). Building on the idea of PAR, the CARP approach brings added dimensions. This holistic and collaborative approach does not just allow the generation of new, locally relevant knowledge and for the uptake of a particular technology. It acknowledges that the realities of farmers are complex, and their challenges are interlinked rather than discrete. Without addressing all the related aspects of land, seed access, agronomic practices, gender, socio-economic issues, ecological issues, market issues and value addition together, the prospects of changing farmers' lives in real and lasting ways are limited. To achieve real transformation in farmers' lives, the

DOI: 10.4324/9781003387497-4

CARP approach integrates knowledge from across different disciplines – from soils, agronomy and plant protection to storage, transport, and then to agricultural entrepreneurship and food processing, food safety and nutrition as well as policy. It facilitates the close collaboration of stakeholders from all sectors and encourages both horizontal and vertical interactions among them. And it helps to develop industries to create better livelihood opportunities for entire communities.

The added value of the CARP approach stems from the bridges of understanding and real-world experience that are built as university research is engaged with, from different angles, by a full complement of stakeholders who all have a unique perspective on the system. The feedback loops work as new learnings emerge from the processes of collaboration and co-creation and give rise to new opportunities for action. For example, as communities undertake value chain activities together and learn from them with different stakeholders, their knowledge of the system grows and so does their ability to spot and successfully develop new livelihood opportunities which emerge as spin-offs while also generating further learnings to inform scientific research. This is not to say the process is perfect or fail-proof – it is built through painstaking and iterative processes of trial and error and building trust.

Rethinking top-down approaches

Several of the CARPs picked up on pre-existing interventions that had been underperforming. In Nakuru County, Kenya, for example, the government had previously attempted to promote the uptake of cassava production in several semi-arid zones located within the larger high-productivity County that otherwise serves as one of the nation's breadbaskets. Farmers in these dry areas were facing season upon season of failed maize harvests resulting from increasing climate variability and diminishing rainfall. Identifying cassava as a drought-tolerant alternative and a potential cash crop, the government provided farmers with planting materials and took the initiative to help them organise into farmer cooperatives but the intervention was not sufficient to mobilise a successful transition to cassava.

Completing their initial baseline survey with farmers, the CARP students and researchers identified several critical communication gaps and missing links that impeded the initiative. They learned that many farmers had abandoned cassava production because of the slow-growing tubers which took up to two years to reach maturity, tying up scarce land resources for untenable periods of time. Crop diseases spread from unclean planting materials and farmers had no means of controlling them. Nor were they equipped to handle issues of toxicity that sometimes surfaced in the cassava. It was also significant that cassava was not part of the local diet and farmers were not well-versed in appropriate varieties for consumption or preparation techniques. Farmers producing cassava struggled to find buyers in the local market. Understanding these issues provided a starting point for researchers to begin working with farmers to establish cassava as a viable crop for food security and cash. Cassava is also critical for resilience to climate change in a vulnerable semi-arid zone.

BOX 2.1 PROFILE OF A PHD STUDENT ENTREPRENEUR

Milcah Wambua, a PhD student in the Egerton University cassava CARP in Kenya, brought a powerful combination of food science expertise and entrepreneurial flair to her work with local farmers. She initially trained the farmers on safe handling and processing of cassava to neutralise the cyanide content which can build up in some cassava varieties under certain growing conditions. She then worked with the farmers to develop a range of different cassava products in cottage industries, including high-protein biscuits, cakes and egg fritters. As she recalls, cassava was an important food security crop where she grew up in eastern Kenya; but as diseases became more prevalent, farmers abandoned cassava in favour of maize. Now reliant on maize, people suffer when the crop fails due to an increasing frequency of drought. She commented:

> What I love about this CARP is that it's all about food security. We have been relying too much on maize, and when we don't have maize there is food insecurity. Cassava does well in drought. I think it will improve food security in the county and create many jobs for people.

Upon completing her master's in food science in 2017, Milcah herself fell into the ranks of jobless graduates. Then she discovered the opportunity to create her own employment by supplying fresh, nutritious, good-quality lunches to lecturers and office workers on the university campus. In her enterprise, she combined healthy, tasty ingredients in innovative ways: traditional Kenyan chapatis made with fresh butternut; pilau stew prepared with meat, spices and fresh coconut added for flavour.

For Milcah, the PhD has gone hand in hand with building her successful food enterprise. She has applied all her combined research, teaching and entrepreneurship skills towards innovating to produce healthy food and contributing towards food security. She commented:

> We have come up with a unique way of putting more nutritious food on the table. We need more alternatives to solve the food issues we face in our society. There are so many opportunities in product development, processing, and value addition. What inspires me is having a country that is food secure, where we don't have issues of children suffering from stunted growth. Without a population that is food secure, we can't grow in other ways. If we are food secure, we will have a productive country.
>
> (Wambua, 2019)

Mobilising farmers

Each of the CARP projects followed an iterative cycle that begins through a dialogue with communities to establish research problems for the researchers and farmers to address together. Through dialogue, researchers and community members began to develop a mutual understanding of the opportunities and challenges of collaboration, and from this, the research agendas took shape. In many cases, the baseline surveys conducted at the inception of the CARPs highlighted clear entry points that provided relatively straightforward tasks to begin working on which would then grow steadily into more complex and elaborate collaborations involving different stakeholders. By addressing these simple entry points first, researchers and farmers were able to remove critical barriers to enable farmers to progress beyond basic subsistence levels. These barriers included the lack of quality seeds and planting materials arising from the poorly developed seed systems in the potato, cassava, rice, pineapple, baobab and safflower value chains, as well as critical challenges of land, water, pest and disease control, that sometimes further stymied production. Without meeting these basic and critical production challenges first, the projects would not succeed in also meeting the equally important but more complex goals of value-addition, agro-processing and market development further down the line.

Many of the CARP projects began with similar starting points and then grew in an organic, step-by-step way to become more complex and involve more partners. In Zimbabwe, for example, water harvesting provided a natural entry point to working with farmers in several neglected dryland and semi-arid locations. Farmers were badly affected by increasingly scarce and erratic rainfall, and food insecurity was growing as harvests failed and families were reduced to eating one or two meals a day. Simple water harvesting adaptations helped farmers to retain moisture in their soils to improve their yields. Long ago under the previous colonial administration, farmers were required to dig channels to divert water away from their fields as an erosion control measure. Repurposing these old structures into 'tied-contour' water retention ponds that then slowly infiltrated the soil enabled farmers to capture rainwater that would otherwise run off their land.

This measure helped farmers to retain up to double the amount of water in their soils, which led to increased yields of maize, which most farmers grew as a staple crop. Next, researchers began working with farmers to replace their maize crops with drought-tolerant and resilient small grains and legumes, including millet, sorghum, groundnuts and sesame that were traditionally grown in some drylands areas and were far better adapted to a dry climate. They worked with farmers to manage their soil fertility and integrate organic manures and small amounts of fertiliser. Taken together, these measures saw farmers' yields increase from 500 kgs per hectare to between one-and-a-half and three-and-a-half tonnes per hectare. Farmers who had previously been producing barely at subsistence level now had surpluses. With surpluses arose the opportunity and need to develop effective ways of storing surpluses and market their grains to buyers. From there, researchers trained farmers in value addition including product development, marketing and creating farmer cooperatives to ensure their development continues beyond the lifetime of the CARP.

From the beginning, the CARPs sought to build an equal partnership between farmers and researchers based on a shared vision, mutual understanding and trust. One key aim of the CARPs was to marry the indigenous knowledge held within communities with the universities' scientific knowledge and capacity for innovation. While traditional research engagements involving indigenous knowledge were often extractive, the CARP design sought ways for these separate knowledge systems to complement one another for mutual benefit. Several of the CARPs focused on indigenous, orphan and neglected crops, including baobabs in Benin, safflower in Botswana, small grains and legumes in Zimbabwe and various indigenous, traditional famine foods in the conflict-affected Darfur region of Sudan. In these projects that have sought innovative ways to utilise previously neglected local knowledge and resources, the strengths of pairing indigenous and scientific knowledge were particularly evident.

In Benin, a multidisciplinary research team developed both cultivation and culinary techniques to germinate young baobab trees and harness their extraordinary nutritional properties to improve nutritional status and livelihoods in communities. Baobabs are one of the most iconic, widely distributed and culturally important tree species found across Africa. Their leaves and fruit contain extraordinary levels of Vitamin C, iron and other key micronutrients. The oil pressed from their seeds can become a highly lucrative export. At least 66 baobab products are already on the market in Europe. Local value chains in baobab remain poorly developed yet have the potential to support new enterprise development and local livelihoods as well as enhance nutrition and food security, biodiversity and climate change adaptation.

The starting point for the baobab CARP in Benin was to develop propagation techniques to make fresh leaves and young trees more widely available as an ongoing food source in rural communities, particularly in the arid northern and central parts of Benin where the trees contribute widely to household nutrition security. Harvested intensively for their leaves and fruit, wild baobabs are threatened by overuse which interferes with their propagation. To address this challenge, researchers successfully developed production from seed to enable farmers to harvest leaves from young baobabs within six weeks and used grafting propagation techniques for earlier fruiting. The research quickly developed into a multi-disciplinary collaboration, as food scientists within the university developed a critical complementary research stream, working with communities to develop and utilise different baobab products such as dried leaf powder and fruit pulp to supplement and enhance the nutritional values of local food products, including porridge for babies, children and women of child-bearing age who are vulnerable to malnutrition.

Ultimately the achievements of the CARPs will only be sustainable if communities continue to benefit in substantial and lasting ways. Innovative thinking is therefore required about how projects can sustain themselves in communities where poverty and access to resources remain critical challenges. In Zimbabwe, for example, new community value-addition centres were built through the CARP collaboration. These centres are stocked with machinery for processing the surpluses of millet, sorghum, *Ziziphus* (masau, a wild tree fruit) and other produce by communities into value-added products to be sold. The centres will need to operate long after the CARP has formally closed. To increase the chances of success, various measures have been taken to promote a sense of ownership and equal access to the facilities. Each family in the community contributed several bricks from which the structures were constructed. To ensure there is money for repairs when machinery breaks down, the communities will invest in keeping cows that can be sold to pay for repairs as the need arises.

One of the most important ways farmers have benefited from the CARPs is by organising themselves and working together to aggregate their resources and outputs to achieve greater economies of scale. Cassava farmers in Kenya, rice and pig farmers in post-conflict northern Uganda, sheep wool farmers in South Africa and others are finding ways to pool their produce to access new market opportunities, develop specialised and lucrative knowledge and skills in seed production, develop cottage industries in agro-processing and more. Traditionally marginalised groups such as women and out-of-work youth are also finding new opportunities to learn, organise and tap into new opportunities. The transformations in communities have started to become visible with the potential to go much further.

Facilitating multi-disciplinary and multi-agency engagements

Each CARP project developed its own systems of collaboration as it grew. The integrated approach of working with communities and developing value chains required researchers to step out of their disciplinary siloes and forge wider linkages within and beyond the university. This required them to become skilled in facilitation and community organising. In Benin, researchers developed relationships with local NGOs and TVETs to carry forward their work on baobab propagation, nutrition and product development and widen their reach in communities. Through the Benin CARP, six value chains of baobab leaves and pulp were mapped, described and analysed, generating evidence to inform entrepreneurs and policymakers. The baobab CARP focused on developing a sustainable value chain for the leaves and fruit pulp, based on the expanded uptake of home garden cultivation techniques that were introduced to relieve the harvesting pressures on wild tree populations. The focus on the leaves and fruit pulp value chains was responding to a need to diversify peoples' sources of income and livelihoods while contributing to meeting food and nutrition security nationally. Ensuring the availability of healthy foods at affordable prices to low-income and marginalised communities is an increasingly pressing challenge not only for Benin but throughout Africa and beyond. Highly nutritious yet under-researched and under-utilised local, wild and indigenous food sources such as baobabs are a critical resource for addressing such challenges. By focusing simultaneously on developing entrepreneurship and livelihoods, improving the quality of local baobab products and addressing food and nutrition insecurity issues, the baobab CARP opened the way for students to develop solutions addressing these interlinked challenges.

Altogether the baobab CARP recorded 34 different local products made from baobab with students working to improve their quality. Gbaguidi Mechak, a PhD student in the baobab CARP focused on developing new methods for producing a traditional fermented drink from the baobab fruit pulp called Mutchayan. Using an alternative fermentation technique known as back-sloping, he focused on improving the quality of the drink, particularly to preserve the naturally high content of antioxidants and Vitamin C. Agossadou Julienne Alawole worked with Professor Flora Chadaré to develop porridge fortified with baobab and moringa leaves. Inspired by the potential of entrepreneurship to address the nutritional status of women and children in Benin, she went on to develop a mushroom business. Students in the baobab CARP both benefited from, and contributed to, a wider coherence among researchers, TVETs, communities, small-scale entrepreneurs, NGOs and the government that has taken shape as the value chain is more clearly understood and the potential of the sector grows clearer.

Often challenges that emerged in CARPs could only be resolved by gathering key people representing different sectors around the table. In the Uganda rice CARP, for example, gathering farmers, rice mill operators and buyers around the table opened avenues for creative problem-solving. Understanding the system better through dialogue, farmers were able to pool their harvests, which in turn made it possible for the mill operators to collect larger volumes of rice and thus provide the critical transport that the farmers lacked. Unlike the farmers, the millers also had the capacity to store the rice and to source buyers. The meetings helped farmers understand and address necessary measures of quality improvement, such as preventing the contamination of rice with small stones during packaging.

For Principal Investigators, the demands of convening dialogues and facilitating collaborative arrangements were intense. They had to convince busy local government officials and other prospective partners of the value of collaboration, which was not always easily demonstrated upfront. In Nakuru County, Kenya, researchers and farmers collaborated in order to kickstart a much-needed local seed potato value chain and address a critical local shortage of clean seed. This required forging strong relationships with a local TVET, the National Potato Council of Kenya, the Nakuru County Government Ministry of Agriculture, the Nakuru Smallholder Farmers Association and the government seed-producing facility ADC-Molo.

Through the project, farmers were trained in methods to produce their own clean seed. It was initially intended that farmers obtain certification as licenced seed growers from Kenya's national seed certification organisation KEPHIS. However, this process was discovered to be overly cumbersome and unaffordable to farmers. The project has assisted them to produce their own clean seed without certification but with demand from local farmers. In the process, the TVET Baraka College and Egerton also developed capacity in seed production and became certified seed potato growers, expanding access to certified seed to small farmers. Although potato is the second-most widespread food crop in Kenya after maize, the potato industry had developed on a largely ad hoc basis and only at the time of the CARP's inception were the policies being developed to achieve coherence in the sector.

This timing provided the opportunity for potato CARP researchers to contribute their technical expertise and key knowledge of the sector as it emerged to inform policy. In the process, they strengthened their relationships with the local government and raised the visibility of the university.

The COVID-19 pandemic posed a challenge to virtually all the projects, as their implementation coincided with lockdowns and distancing measures. In many cases, field trials were interrupted so that farmers and researchers could not meet in person. On the one hand, the pandemic posed challenges to the hard-won cohesion that was beginning to emerge among stakeholders as in-person meetings and travelling to the field became increasingly difficult. On the other hand, farmers and researchers working on some of the projects also learned that they could still achieve important progress and sustain their activities using social media, WhatsApp and Zoom meetings.

One thing leads to another

Most of the CARPs were characterised by a growth pattern of 'one thing leading to another'. Several of the Principal Investigators and researchers associated with various CARPs have come to recognise that the 'real research' begins with community engagement. As projects developed momentum, they grew in complexity and new previously unseen avenues for development emerged. As stakeholders understood the different facets of their value chains better, they could more easily recognise new opportunities for innovation and growth. In most of the CARPs, momentum grew through stakeholders in the value chain organising themselves, building relationships, seeing more of the 'whole' of the value chain and involving diverse perspectives. This opened the way for new innovations in product development and value addition.

In the Uganda rice CARP, a student from Benin, a small West African nation where rice growing is more established and prevalent than in East Africa, was able to advise farmers on techniques for parboiling rice as a value-added product. The student showed platform participants YouTube videos of the process in her home country. From this interaction, TVET students were able to fabricate large rice-parboiling drums from local materials in Uganda.

Clean-burning charcoal briquettes were another business spinoff that emerged out of the awareness of a growing waste problem at the rice mills. Behind every mill, there lay a growing mountain of discarded rice husks. The rice mill operators did not know what to do with them, but students saw an opportunity to turn 'waste' into wealth. Taking the problem as inspiration, students developed a formula for producing charcoal briquettes using the discarded rice husks to produce a clean-burning alternative to wood charcoal which is used as a fuel source by most of the local population and is a major contributor to deforestation. The CARP partnered with the Uganda Industrial Research Institute to develop the product and receive assistance with designing the briquette mould and the equipment for carbonising the rice husks to yield a high-density energy source.

The CARPs in cassava, rice, potatoes, baobabs, safflower and others have similarly reached stages of commercialising diverse new products and catalysing small-scale farmer industries.

BOX 2.2 MARRYING SCIENCE AND INDIGENOUS KNOWLEDGE SYSTEMS

Zimbabwe CARP Principal Investigator Ronald Mandumbu (2022) remarked:

> As researchers, we no longer have the ivory tower mentality that because I come from the university, I know everything. In the process of research and development of the products, you are humbled by the communities when they suggest things that you have never imagined as an intellectual. When I went first into the community I thought I was going to teach them. I had the ivory tower mentality, until I listened to the communities. You discover that they have their own knowledge systems. Their indigenous knowledge systems are rich enough to be translated into tangible products. They might need our technical know-how to perfect certain things, but their practices have been in use for a long time. We asked the old ladies, how would you prepare this? We saw the nutrition, the combination of the ingredients they used was surprising. When we checked the nutritional content of all the products that were put in the Rupiza products (the cowpeas and peanut butter soup) we said this is superb. It's like they were nutritionists who knew there is something in peanuts and sesame, something in cowpeas and we mix the ingredients, roast them and produce a product. That is exactly what they do in industry. As a researcher you are humbled by what the community can teach you. You respect the community because it is a source of information for you. You are going to learn.

Such processes of discovering how to utilise and get more value from local products that were previously overlooked have been important to the CARPs. In Zimbabwe prior to the CARP, for example, there was no local market for sesame, which is viable in dryland areas. However, the only option for farmers who produced sesame was to sell it cheaply to cross-border traders from Mozambique. With the CARP project, a spate of product development was catalysed in the university which has begun providing new outlets for sesame and other previously neglected produce.

Interviewing older women in the community, researchers discovered the abundance of a local wild fruit called *Zizyphus mauritiana* (locally known as 'masau') growing prolifically in the local landscapes. The elders recalled that in past years of scarcity and drought, they would pick and dry the fruit, remove the seeds, and mould it into cakes to sustain community members when all other food sources were exhausted. This insight provided the inspiration for developing a modern energy bar mixing masau with groundnuts, sesame and honey to provide a convenient and nutrient-dense food source which could be distributed in schools as well as sold commercially. In this way, new avenues for innovation and product development continue to open.

Strengthening systems, tapping new opportunities

As the universities have gained the ability to broker partnerships, these partnerships have improved the visibility and effectiveness of all the platform members, from the universities to marginalised and remote rural communities. Through making valuable contributions in areas of policy and industry, the universities have been extending their spheres of influence and their attractiveness as partners – a theme that will be elaborated in the following chapter.

Meanwhile, farmers have learned that they can diversify their activities, organise themselves and develop more lucrative opportunities in value chains, including seed production and agro-processing. As the stakeholder platforms have become more established, farmers have steadily gained the ability to broker new arrangements and expand their activities. The universities would not have been able to catalyse the same extent of transformation acting alone. Through partnerships, wicked problems have been addressed through multifaceted interventions. The momentum and visibility of the platforms have begun to pique new interest from other potential partners that can help take certain products to a wider scale. The pineapple CARP in Ghana has triggered businesses in planting materials and information using drones and social media. The wool CARP in South Africa grouped farmers and linked them to wool export markets. It also trained and supported handicraft production and marketing. The cassava CARP at the University of Nairobi and the pig CARP at Gulu University significantly improved productivity and strengthened community and student entrepreneurial skills and businesses. Largely through the catalytic efforts of the CARP, Benin now has more than 163 small enterprises working on the development of the baobab value chain. Former CARP students have graduated to become entrepreneurs and employ others and contribute further to the development of the ecosystem.

Through the CARP facilitation, challenging processes such as farmer seed certification have been navigated which have demonstrated the value of the approach to farmers. Through the rice CARP partnership in Uganda, for example, farmers were able to obtain certification of the quality of their seeds from the Ministry of Agriculture, an important step towards earning more money for their seed. To be certified to produce and trade seed would have been prohibitively challenging to an individual farmer, but when farmers formed groups through the partnership it was feasible.

Such measures have improved the attractiveness of farmer groups to other farmers. Other farmer groups have succeeded in developing proposals and attracting government support. One group mobilised support to buy a tractor. Another cooperative raised close to US$10,000 (350 million Uganda shillings) to build a store and equipment to mill and package their own rice. Other farmers have done their own packaging, brand, logo and quality mark so they can trade their own higher quality rice. It will be visible and traceable in the market.

In Botswana, the CARP in safflower has enabled farmers to organise themselves to form cooperatives and proactively drive the growth of the sector. A neglected and unknown crop in Botswana, safflower has extraordinary potential as a high-value dryland crop with strong health benefits and widely diverse applications including silage, hay and seed or cake meal for animals, fresh green vegetables, cut flowers, healthy high-quality cooking oil and natural pharmaceuticals from processed dried petals. Forming cooperatives, farmers have enjoyed easier access to finance and inputs and even imported their own seed when local stocks were inadequate. Farmers have successfully mobilised through the CARP to sell numerous products including fresh greens to supermarkets and ground petals for tea infusions and animal feed which has already had an impact in reducing the nation's import duties on animal feed imports. Women who had suffered gender-based violence during the COVID-19 lockdown were trained in safflower production, product development and agribusiness through the CARP and subsequently secured a tender to supply fresh green vegetables every week to a school with 3,500 children. The women registered a cooperative called 'Green Diamonds of Botswana' and in April 2023 planted 20 hectares of safflower. Already in the short life of the project, drylands communities have seen marked improvements to their food security, income levels and livelihoods through safflower production.

An area where farmers have fallen short in their aspirations, however, is the production of highly valuable safflower oil, which requires high volumes of seed production and substantial investments in industrial machinery and infrastructure. Farmers have lobbied their local Members of Parliament for new policies and incentives to support the growth of a domestic safflower industry. In response to the demand of farmers, and because of the success demonstrated by the CARP, the Botswana Government has piloted in one district the initial steps to support the growth of an oil industry. In the pilot district, the government has stipulated that at least one of the 4 hectares per farmer of land ploughed by the government, must be planted with safflower.

The CARP has been well aligned with the Botswana Government's quest to promote household food security by supporting micro-scale farmers and enabling them to contribute significantly to household food consumption needs through the provision of inputs, finance, rain-fed production systems and other measures. Safflower was among the 13 crops included in and supported by a programme 'Temo-Letlotlo' which began in April 2023 with the goal to promote production of cooking oil from safflower and sunflower in order to reduce cooking oil imports. Under the programme, the government will give subsidies to all farmers in the country to produce and process safflower, which is likely to contribute further to improving food security and nutrition.

Although the university was the catalyst for safflower production in Botswana, it is now the farmers driving the growth of the industry. They have benefited from the high demand for safflower products and seeds. The project has taken on a life of its own through the initiative of the farmers.

As several CARP Principal Investigators have observed, the failure of well-developed value chains in orphan and neglected crop areas has been no accident. It resulted from a lack of focused investment in research and development. In many rural communities, elderly women have been the custodians of traditional varieties of sorghum and millet, for example, that would have been appreciated as community and cultural heritage but are no longer farmed by younger generations. While indigenous knowledge systems of production of traditional crops have been steadily dying out with older generations in rural communities, indigenous and wild foods, as well as neglected and under-utilised crops, are at the same time enjoying a revival of interest as their nutrition, health and climate resilience attributes become more widely appreciated (Searchinger et al., 2019; Snapp et al., 2018; Waha et al., 2018).

Several of the CARPs have highlighted a relationship between scientific innovation and the resourcefulness of utilising unique local knowledge and products. In the face of climate change, these are becoming critical resources for adaptation and resilience. The famine food CARP in Sudan highlights how important it is to conserve and adapt traditional local knowledge. The Dande Valley in Zimbabwe where the CARP platform engaged is also dealing with severe climate-related stresses and shocks. The ingredients used in the products developed through the CARP, including *Zizyphus mauritiana* porridge, are all drought tolerant and climate resilient, produced by drought-tolerant varieties and grown on wild trees with deep root systems.

The CARP from the University of Khartoum in Sudan focused on identifying and promoting survival strategies in the drought and conflict-affected Darfur region, based on indigenous, wild and fermented food sources. Engaging with the extreme challenges of Darfur, the work is deeply significant both in terms of helping communities affected by famine and of producing critical knowledge of little-known plants with a huge relevance for biodiversity, agrobiodiversity and human nutrition. The project exemplifies how many of the CARPs engage in response to key local issues on the ground, yet at the same time produce knowledge with wider regional and even global developmental significance.

The Sudanese project documented the foods consumed in times of famine before relief efforts could reach affected areas and gathered knowledge of the nutritive values and preparation techniques used. Researchers and students worked with primarily female farmers as well as owners of small-scale fermented food businesses. Through the project, researchers and farmers were able to identify products with higher nutritive values that were well-adapted to the environment, such as fermented hibiscus seeds known as furundu and highly nutritious fermented locusts known as fendo. The researchers drew on science to supplement and improve these products as a strategy to improve flexibility and food security in Darfur. Limited in scope and by the conflict, the project only scratched the surface of a vast wealth of plants and indigenous knowledge important for survival in vast areas of Darfur and elsewhere in Africa. Yet it points to a critical role that universities can play in harnessing science to support the resilience and adaptation of local knowledge and support communities facing extreme survival challenges.

Conclusion

In all the settings where they were piloted, the CARPs catalysed diverse small-scale industries and enterprises. This was achieved through working closely with communities and building collaborations among different value chain stakeholders in local industries, NGOs and governments, TVETs and extension services. The ground-up, iterative nature of developing the platforms has generated momentum to assure the projects' sustainability and continued growth. In several projects, including the Uganda rice and Botswana safflower CARPs, the evolution of small farmer groups into larger cooperatives with increasing capacities to act on their own initiatives has helped to support the longer-term sustainability of the work. In several of the universities, policies on outreach, community engagement and innovation and industrialisation provide another measure of institutional support to facilitate the continued engagements of the platforms. In Benin, the CARP focused on capacity building to produce a critical mass of well-trained students as a sustainability measure to support the continued development of the baobab industry. In Zimbabwe, the CARP enabled scholarships to be provided for the training of several unemployed youth in the community at the TVETs as an additional investment in the future of the community.

All of these different elements are critical for the sustainability of the multi-stakeholder approach that was nurtured through the CARPs, but none is sufficient on its own. All the different elements cannot work in isolation.

The CARP approach has nurtured collaborative processes of community-based agro-industrialisation. Working with researchers and students, as well as with a wide range of national and government agencies, civil society organisations and the private sector, farmer groups have begun to overcome the barriers that kept them locked in poverty, and in some cases have organised through cooperative structures to become stronger advocates for their continued development. This chapter has illustrated the importance of building collaborations and multi-stakeholder platforms facilitated by universities that support the different facets of community-based agro-industrialisation.

Together the CARPs have demonstrated a diversity of value chain activities undertaken in diverse contexts to support increased farmer and community livelihoods. Many practical enterprises have stemmed from knowledge co-production. The platforms have linked communities and the students with research and development agencies, relevant civil society or government agencies, entrepreneurs and marketing and information systems that support the commercialisation of smallholder production. Collaborations between indigenous knowledge and science have spurred innovation and knowledge to be utilised by communities. Through these processes and partnerships, universities have developed their capacities to catalyse and facilitate multi-sector, multi-stakeholder innovation platform engagements. In the process, the CARPs have helped to strengthen the capacities of universities, communities and other actors to apply science towards developing new industries that provide the greatest potential to lift rural people out of poverty.

Bibliography

Adipala, Ekwamu, Frans Swanepoel, Melody Mentz-Coetzee, Farai Kapfudzaruwa, and Kay Muir Leresche. 2021. 'Agriculture and Tertiary Education Within the Context of Global and African Development Goals'. In *Transforming Tertiary Agricultural Education in Africa*, eds. David Kraybill, John Lynam and Ekwamu Adipala, 3–27. CABI.

Juma, C. 2012. 'Building New Agricultural Universities in Africa'. Faculty research working paper series. John F. Kennedy School of Government. 12(26).

Kibwika, P. 2006. *Learning to Make Change: Developing Innovation Competence for Recreating the African University of the 21st Century.* Wageningen Academic Publishers.

Mandumbu, R. 2022. Personal Interview.

Mponela, P., Manda, J., Kinyua, M., and Kihara, J. 2023. 'Participatory Action Research, Social Networks, and Gender Influence Soil Fertility Management in Tanzania'. *Systemic Practice and Action Research*, 36(1): 141–163.

RUFORUM. n.d. 'Engaging Universities to Strengthen Agricultural Value Chains'. Video. https://ruforum.org/impact/project/engaging-universities/.

Searchinger, T., Waite, R., Hanson, C., and Ranganathan, J. 2019. 'Creating a Sustainable Food Future: A Menu of Solutions to Feed Nearly 10 Billion People by 2050'. Washington, DC: World Resources Institute, July 2019. https://www.wri.org/research/creating-sustainable-food-future.

Snapp, S., Rogé, P., Okori, P., Chikowo, R., Peter, B., and Messina, J. 2018. 'Perennial Grains for Africa: Possibility or Pipedream?'. *Experimental Agriculture*, 55(Special Issue 2): 251–272. DOI:10.1017/S0014479718000066.

Tizikara, C., Nampala, P., Nielson, D., Beintema, N., Okori, P., and Lynam, J. 2021. 'Transformative Research and Innovation Capacity in Tertiary Agricultural Education in

Africa'. In *Transforming Tertiary Agricultural Education in Africa*, eds. David Kraybill, John Lynam, and Ekwamu Adipala, 135–155. CABI, Oxfordshire.
Waha, K., van Wijk, M. T., Fritz, S., See, L., Thornton, P. K., Wichern, J., and Herrero, M. 2018. 'Agricultural Diversification as an Important Strategy for Achieving Food Security in Africa'. *Glob Change Biol.*, 24: 3390–3400. https://doi.org/10.1111/gcb.14158.
Wambua, M. 2019. Personal Interview. Egerton University, Kenya.

3

UNIVERSITIES FIT FOR PURPOSE

Megan Lindow, Anthony Egeru and Kay Muir Leresche

Introduction

Change processes require leadership and intention and often encounter resistance along the way. Each of the community action research projects (CARPs) outlined in this book provided a lever for change within institutions and helped them to grow more 'fit for purpose'. Mindset changes were often key to these efforts, as researchers and students learned to see themselves as change agents as much as scientists and prioritise roles as action-oriented facilitators and organisers. The CARP engagements helped to build institutional capabilities and practices in diverse ways, albeit incrementally and on a small scale. The PhD and master's scholarships provided through the CARPs helped build the capacities of younger academics to bring a full complement of diverse skills to institutions to implement in the classroom and in their research. The relationships built with communities and other stakeholders through the platforms generated new opportunities for working together. The diverse experiences of the CARPs provided useful and relevant material to rejuvenate curricula in the universities and TVETs, further extending the learnings of the platforms. By stimulating value chain development in areas critically important to local communities, such as indigenous foods, dryland crops and key food security crops such as potatoes and cassava, the universities wove themselves into these key innovation systems, contributing to the strengthening of the whole fabric. Through these engagements, the universities had to develop their own resources in thought leadership and facilitation of change, advancing the value proposition that academics and universities bring to the table.

A key conversation is about shifting universities from a model of Education 3.0 (teaching, research and outreach) to Education 5.0 which adds innovation and industrialisation. African university libraries are full of dissertations and business

DOI: 10.4324/9781003387497-5

plans which require translation to products and services in a more meaningful and dignified manner. The idea of Education 5.0 is that universities must also be actively producing and testing products and services with communities to experience how they work in real life. This model takes students through the practical steps of research, product development, testing and commercialisation. As students acquire such practical experience and skills, they develop as the entrepreneurs and job creators that Africa so desperately needs. Such engaged and practical learning modalities help further to attract employers and multi-sectoral partners to the universities.

Numerous research and community initiatives produced through the CARPs over the last decade led to new products and services developed by students, communities, or both, working together. In Benin, student entrepreneurs developed nutrition-focused businesses in the baobab value chain. In the South Africa wool CARP, students developed spinoff businesses in offcut felt wool jewellery as well as in agricultural advisory and extension services. All of the current CARPs covered in this book have resulted in multiple student and/or community enterprises. These are just a few illustrations of universities in motion shifting the narrative as development facilitators. These and other spinoffs of the CARPs are covered more fully in Chapter 5.

Above all, moving the universities from theory to practice requires providing a policy environment and an institutional culture that encourages and supports practical work. Numerous lessons both large and small surfaced through the CARP engagements. The experiences highlighted university policies and practices in procurement that are sometimes not well aligned to the demands of working with communities. In terms of policy, many universities have yet to update their policy guidelines to reflect the higher value aspirations of more enterprising Education 5.0 approaches. Accordingly, leaders of the CARPs have argued that new policies are needed to guide how academic promotion processes work, for example, those that put greater emphasis on community engagement and mentorship of student enterprises, with less weight attached to journal publications. Increasingly, universities are finding ways to reorient their policies and cultures towards supporting practical learning and research through community engagement, innovation and agro-industrialisation. Gradually, universities are recognising a need to re-interpret the narrowly conceived roles of faculty in their human resource manuals, just as they are waking up to the disconnect between their often outdated and overly theoretical curricula and the complex skills and knowledge demanded in the world their graduates will be entering (Kraybill et al., 2021).

Bridging the siloes

While the last chapter included themes of interdisciplinarity from the perspective of building farmer collaborations, the issue also deserves attention from an institutional point of view. Trained in their disciplinary siloes, academics have often found it challenging to adapt interdisciplinary and transdisciplinary approaches. Across the CARPs, however, looking beyond traditional disciplinary boundaries produced new learnings and levels of insight. In the Uganda potato CARP, for instance, researchers were well aware of the multi-disciplinary challenges that limited both crop production and market access. As an agricultural economist working on potatoes since 2002, the Principal Investigator Professor Johnny Mugisha could see the disconnects that impeded development of the value chain. He could see that a multi-disciplinary approach was required to address these challenges. Farmers in densely populated western Uganda were severely limited in land access. Potato yields were falling, exacerbated by farmers using their smallest tubers as seeds, meaning they were effectively selecting and recycling the lowest quality tubers. There was almost no value addition and potatoes were sold raw with no potato products developed and marketed. The quality of potatoes produced did not meet the standards of fast-food companies which then resorted to importing potatoes.

The challenge of the CARP was to help transform this situation through a connected and multi-disciplinary effort. Addressing land use and soil fertility, seed, agricultural production practices, value addition and marketing issues together required a different orientation to more conventional research. Addressing any of these components on its own would not help the overall situation for farmers. Simultaneously, the farmers needed training in seed multiplication and crop intensification to help them make the most of limited land resources. Only then would they benefit from value addition.

For Mugisha, bridging traditional academic siloes brought new opportunities for fresh thinking to the project. Confined to home during the COVID-19 lockdown in 2020, he noticed that the household potato peelings were sprouting into new plantlets in the compost pile. That brought forth the memory from his own childhood of how his parents would make their own selections of good tubers from previous harvests. They would let the tubers sprout and then cut them into two or three pieces and plant the cuttings. The practice had long since disappeared, but the memory gave Mugisha the idea to assign a student to experiment with using cuttings as seed, to enable farmers to stretch their limited seed supplies further.

The project needed a food scientist, and a master's student was recruited to work on value addition. As Mugisha recalls, the student's supervisor wanted him to work on developing French fries.

BOX 3.1 TAKING A MULTI-DISCIPLINARY APPROACH

Mugisha (2021) recalls:

> As a Principal Investigator, I was not excited. How will French fries (chips) help the farming communities? The farmers in the rural areas don't eat French fries. Only a very small top class of people in urban areas eat French fries, so what help am I giving to the farmers?
>
> The food scientists said, but what else can we make out of potatoes? I suggested we try to process the tubers into potato flour, but they said it was

impossible. We were going to the field in one car, and I was trying to argue with them. I said, you're the experts, but I have a conviction that it's possible. We stopped the argument there, but I talked quietly to the graduate student. I said, for your dissertation you can work on French fries, but I'm not excited about it. Let's try and see if we can make flour.

The student was very innovative and enterprising, wanting to try every idea that came his way. He peeled the tubers, pounded the potatoes and dried them out in the sun. He produced some flour, baked some cookies and brought them to me. That was the most exciting part of the project. The cookie was crooked in shape, and one side was raw while the other was burnt. But it was possible. From there we improved the processes and came up with marketable cookies, biscuits and waffles. The flour can be fortified and mixed with other nutrients so that it becomes a complete food product. These products convert what would otherwise be low-value farm waste, because the small tubers are being used, so we are adding value without taking away from the market for the bigger tubers. The food science student has teamed up with an agribusiness student and formed a company, the Muhingi Products Ltd. They are taking the products to market.

As the story demonstrates, the multi-disciplinary platform approach made it possible for students and researchers to think outside the box, make connections and explore unconventional routes that lead to novel problem-solving. To do this, they had to step around conventional patterns of thinking within the institution. Conventional, siloed research can tend to rely on established practice (i.e. potatoes are used to make chips). It does not often prioritise coming to a deep understanding of the challenges facing communities. In contrast, the iterative and ongoing nature of participatory research platforms provide a structure that encourages experimentation and innovation as part of the collaborative learning process. The step-by-step approach of the CARPs outlined in Chapter 2 opens up a wider view of what can be done and has therefore helped to build momentum in the projects. This collaborative, holistic, cross-disciplinary and multi-agency aspect of the CARPs has been essential to enabling movement and change in the lives of communities. It has also helped universities to experience and appreciate rural development as part of their mission.

By reorienting towards multi-stakeholder research through the CARP, the university has provided a collaborative platform for students to gain real-world experience and the practical skills and confidence to become entrepreneurs and develop networks. For example, the student who worked on seed multiplication in the

Uganda potato CARP now plans to become a commercial seed grower, mirroring students and farmer groups from other CARPs including rice, baobabs, pineapple and cassava who have also initiated businesses producing quality seeds and clean planting materials. Another spinoff of the CARP, the Muhingi Products Ltd. (see Box 3.1), is a business created by students that produces potato-based products, employs 11 people and sources their potatoes from 100 farmers from the community. When Mugisha presented the potato CARP project to the university Vice-Chancellor, the leader challenged him to use the project to build a potato processing factory in the community, rather than at the university. Mugisha (2021) reflected:

> That made me feel good, because typically if you are given resources and equipment through a project, the university policy is that when the project ends the equipment belongs to the university. But the Vice-Chancellor instead said, take the equipment to Kabale where the farmers are and start processing potatoes there. He understood the value of this direct engagement with farming communities for the university.

Building synergies between universities and TVETs

By design, the CARPs brought the universities and TVETs (technical and vocational colleges) closer together and enabled them to form lasting relationships. Like the universities, the TVETs suffered long periods of underfunding and neglect from the 1980s onward. Historically the universities and TVETs had little to do with each other, with agricultural TVETs often reporting to the Ministry of Agriculture rather than to a ministry of education. Yet the two types of institution complement one another. The universities provide general professional degree training and supply teachers and lecturers for the educational pipeline, in addition to driving research and innovation. The TVETs carry the mandate to train students to apply the practical applications derived from the universities' research and development. TVETs are usually much more closely aligned with communities on the ground and also provide lifelong learning opportunities for extension and other community knowledge workers and farmers.

Key to transforming institutions to be 'fit for purpose' is to ensure they are positioned to deal with the youth unemployment crisis and are agile enough to respond to emerging challenges. Here, the TVETs have an essential role to play, particularly because of their practical orientation and proximity to communities. The CARPs enable relationships between the universities and the TVETs to develop and grow, based on practical engagement. The TVETs offer a range of diploma and certificate programmes, training technicians for specific fields but also, depending on the particular programme, teaching a broad array of practical skills such as welding and machinery operation that enable the wider practical uptake of university-led innovations on the ground with communities. Through the CARP platforms, the TVETs were able to train young people with relevant skills in value chains. In many cases, TVET students fabricated equipment used by farmers on projects and TVET institutions provided land for participatory field trial and training sites which brought together communities and other platform stakeholders. Through the CARPs, several of the TVETs were able to upgrade their curricula to include new innovations, while the knowledge they acquired allowed them to make more relevant contributions within value chains. Importantly, the TVETs became centres for last-mile delivery of technologies, innovations and management practices. At Baraka Agricultural College in Kenya, for example, newly developed capacities in potato seed multiplication led to increased dissemination and access to quality potato seed for farmers in Molo, Nakuru and Nyandarua, the key potato growing areas in Kenya.

BOX 3.2 A UNIVERSITY AND TVET COLLABORATION IN UGANDA

Through the Uganda pig CARP, technologies were developed from local resources to help farmers improve the quality of their pig production and tap into the growing consumer demand for pork products. Pigs reproduce quickly and can be fed from diverse sources including crop residues and agricultural waste, making pork an ideal protein source in regions such as Gulu that are impacted by poverty, population growth and climate change. However, before the CARP intervention, only 20% of farmers surveyed were rearing pigs due to numerous challenges: they lacked proper food formulations and struggled to manage diseases and maintain sanitary conditions for their pigs (RUFORUM, 2021).

Bringing Gulu University together with the Northern Uganda Youth Development Centre, a local TVET, and farmer organisations, the local government, national veterinary advisory services and private pig and pork trading and marketing associations, the CARP sought to address these barriers to the growth of the pig value chain. Among the innovations developed through the CARP was an artificial insemination (AI) technique, utilising local coconut water as a semen extender, to boost the size and health of pig litter and prevent the spread of disease. The researchers also assessed and developed feed formulations from such locally available resources as sweet potato vines, cassava tubers and rice bran to give farmers affordable feed sources without compromising the health of the pigs or the quality of the meat. To address pervasive problems of foul odours from the pigs, researchers delved into the global scientific literature and adapted a solution from Korea which involved constructing pig houses with deep litter floors treated with indigenous micro-organisms (IMOs).

Young people from the TVET were trained alongside other community-based organisations to enable them to continue to train youth and farmers.

This opened new opportunities for youth to provide the AI and IMO services. The TVET also then went on to offer technical assistance in both AI and IMO technologies as paid services to local farmers. These interventions have built practical knowledge as well as social cohesion across the various stakeholders in the platform, particularly through the labour-intensive process of constructing the pig houses, which requires community members to work together and support one another. As the report from one training meeting described: 'Sharing the research findings brought more joy, love and hope for agriculture to the participants and also the Gulu and TVET students because they practically participated in setting up the experimental unit, trapping the IMO, making the IMO solution and laying the deep litter floor IMO bed' (Lindow, 2020).

The power of building these social relationships through practical work ensures that young people find new business opportunities in the value chain. As the universities produce such new technologies and innovations as the AI and IMO, providing locally affordable and appropriate solutions to farmers who could not afford more 'conventional' solutions, they have an equally important opportunity to link with the TVETs to harness that capacity to train young people in providing the services that will allow wider uptake of the technologies among farmers.

In the Zimbabwe CARP, some 34 young people (several of whom were selected from communities such as the Dande Valley where the participatory research took place) were trained in a local TVET as part of the project. This was done to strengthen relationships with the community. It also provided a lasting footprint as the students would be empowered with skills to continue the work of product development and market access in the community after the end of the project.

Training local young people in the TVETs has relieved a burden on the community. Some 80% of youth remain unemployed in Zimbabwe. The problem is particularly acute in the Dande Valley, a marginalised drylands community with few job opportunities outside of subsistence farming, where many unemployed young people continue to languish without skills or prospects for further studies. The hopelessness of their situation breeds anger, discontent and drug use. 'We have seen that youth do not need to have five "O Levels" to have the capacity to be trained and gain skills. If we are to deal with youth unemployment decisively, young people need to be in a system that allows them to gain skills. Once somebody gains skills, they have the capacity to start new things and find opportunities', observes Zimbabwe CARP Principal Investigator Professor Ronald Mandumbu (2022). The realisation that the community has valuable resources such as *masau*, a wild indigenous fruit, and sesame that can be developed and commercialised has also been exciting for young people. If the products developed through the CARP can be successfully commercialised, this will give the community further resources to invest in training young people through the universities and TVETs.

Meanwhile, in the Kenya potato CARP, the participation of the Baraka Agricultural College, a local TVET, has transformed training and engagement. Potatoes are a US$500 million industry for Kenya and the second most important food crop. Yet potatoes have also been a neglected area of research investment, as the sector arose informally and previous policy and donor involvements favoured maize. As such, potatoes were an impactful area for the universities and TVETs alike to engage with.

As Baraka College underwent certification as a potato seed grower, TVET students participated fully in the process and gained experience in all aspects of seed potato production. The engagement strongly impacted teaching and practical learning, as the College developed a modern seed storage facility and a demonstration plot to integrate and showcase diverse production practices among students and farmers. Through these interactions, the TVET was able to integrate deep and diverse scientific knowledge with practical experiential learning and training to make these accessible to farmers and communities. 'The niche of Baraka College is to deal with small-scale farmers, so that they can get the skills to put food on the table and make money. The fact that we were not growing potatoes here before was a mistake', remarked former Baraka College Principal Virginia Nyamu (Lindow, 2020).

Institutional mindsets and culture change

By design, the cultures of many African universities built on the models of colonialism and the ivory tower have been bureaucratic, conservative, resistant to change and male-dominated (Kraybill et al., 2021). The demands of responding to complex challenges of food security, youth unemployment, rural livelihoods and climate change are putting universities under intense pressure to adapt rapidly. Across institutions, the CARPs sparked important mindset shifts in different ways to support the role of universities as development facilitators and partners engaging at the local level with national, regional and global priorities. A key lesson learned was that in order to catalyse the kinds of societal transformations they seek, universities must in turn transform themselves. The CARP approach provided a lever for transformation, albeit often on a small or limited scale. Change efforts sometimes incurred resistance, as Principal Investigators, researchers and students involved in the CARPs encountered barriers of fixed and entrenched institutional practices which at times impeded their community-building efforts. Yet, through such sticking points, the CARPs highlighted important areas for universities to transform themselves to deepen their relevance as fulcrums for development.

The CARPs highlighted areas where universities needed to refine their systems and processes to support those outcomes described above which demonstrated their relevance. In some cases, researchers faced frustrating delays with procurement and the disbursal of funds, because the policies of their institutions were not attuned to the time-sensitive demands of community engagement.

BOX 3.3 CONFRONTING BARRIERS OF INSTITUTIONAL CULTURE

One Principal Investigator remarked:

> A negative trend I have seen across institutions is that administrators don't understand what we are doing as researchers in agriculture. When I say I am taking all my students and team members to the field to train 500 farmers for one week, and we need two vehicles, their response is, What? You're going to blow all this money on that? But I am not blowing money, these are necessary activities. This is something that the finance people and administrators and sometimes even the vice-chancellors don't understand.
>
> Or you raise a claim and even after three to six months it is not approved. But time is moving and agriculture is seasonal. Planting must be done when it is time to plant. So you end up failing to complete a planned activity because the money was not released in time. We have very good relationships with the community. If we tell them we are coming on Wednesday to plant, we get there and they are waiting for us. They have stopped other important activities to be there for us. So if you don't get your payments to travel, you look like you are not trustworthy and this begins to damage your relationships with the community.
>
> Another time we went to the field without getting the money – it had rained, we had to plant. Fortunately, we had the inputs, but we did not have travel money or money to pay for labour in the field. These farmers took their manure and used it, and we have not paid for it. They took their cows and oxen and helped us for three days. It's a challenge.
>
> (Anonymous, 2021)

Most of the CARPs faced similar challenges. Yet as the CARPs began to demonstrate the value of action research, entrepreneurial and experiential learning and community engagement to the wider relevance of the universities, some administrators became more accommodating. In this way, small changes sometimes filtered up to the levels of university leadership and seeded new mindsets and approaches.

In Sudan, where four smaller CARPs operated, previous engagements between the universities and communities were extremely limited. The CARPs had the effect of stimulating an engaged, experiential and collaborative style of research in universities where this had rarely existed before. While most of the universities where CARPs operated had to navigate unfamiliar challenges brought by an experiential approach not yet embedded in institutional practice and culture, in Sudan the novelty of the approach was even more pronounced. Sudan's university system is extensive but has stagnated and been isolated from the rest of the world due to long-standing political and economic turmoil, including a revolution in 2019. Such common challenges in African universities as outdated and overly theoretical curricula and jobless graduates are even more pronounced in Sudan. In this context, the strength of mentoring and support provided by RUFORUM as a wider institutional network was pivotal to all four Sudanese CARPs, which were transformative for their universities.

BOX 3.4 COMMUNITY ACTION RESEARCH WITH RURAL WOMEN IN SUDAN

The University of Gezira, in the centre of Sudan, had produced a number of research findings but had never run a project that engaged the community. As Principal Investigator Professor Muna Mohamed Elhag describes it, initiating a modest project with local women facing food insecurity and dependent on their husbands' meagre incomes as migrant labourers was a process of finding their way. The project involved working with the women to introduce home vegetable gardens and integrate small livestock rearing, in order to provide the women with household food security and a supplement to their incomes by selling their produce.

Rural Sudan was suffering intensely from the country's political unrest and economic collapse, combined with the effects of climate change which have led to major losses of the staple sorghum crop. In the past, villagers could always dip into their sorghum stores if they needed money. But now, due to all these hardships, the sorghum stores are depleted. Building their engagement with women from the local community, researchers and students had to learn to interact very differently to how they operated in an academic culture. Long hours had to be spent drinking coffee and visiting the homes of community women before work in the field could progress. This required an attitude of openness and willingness to let the process develop organically without rushing ahead. Once the researchers got to know the local women, they gradually brought seeds and started to work with them to prepare the land and introduce new crops, chickens and goats.

Two students spent six months living in the village, and the time was transformative. They worked with the women to build greenhouses and chicken houses, prepare compost and collect data. One student found that the nutrition status of children under five has improved. New observations led to new interventions as the project unfolded. The villagers were enthusiastic about raising chickens, but these were vulnerable to disease, so the students helped to bring vaccines to the village and also introduced more disease-resilient quails. They also noticed the village had no fruit trees and began planting them. Young boys were particularly enthusiastic about the fruit trees.

Muna Elhag says: 'What surprised me was that boys of 8 or 10 wanted to plant their own trees. If we can encourage them to think like this, we will have a new generation that is very linked with agriculture' (Elhag, 2022).

Muna Elhag (2021) reflects:

> The programme has changed a lot in the village, but there is still a long way to go. We have brought the rural people to the university. They are so happy. They gain knowledge so fast. They start to see that academics can help them get information. Sometimes they call me to get advice about the best variety, or because they have found some disease. We already managed to develop a module in chicken management and manure compost production that we can take to the community and train them. This project has been a good way to make a pathway between the university and the community.

Muna Elhag (2021) continues:

> My university started in 1977 and the curriculum has been here for a long time. I think a programme like ours can change the way we think about the curriculum. It can be transformed to meet the requirements of the community. There are no gender issues in the curriculum. There is nothing about climate smart agriculture, and nothing about all the new technologies that can help communities. All of this can be changed through a programme like ours. We need to change our way of teaching the students, so we are not just giving them scientific facts. We need to mix facts with real experience from the field. We need to go to the village and see what climate smart agriculture means for the people. We need to have this linkage between real life and scientific facts.

Becoming fulcrums for development

SDG 17 realises the importance of partnerships for meeting all the other Sustainable Development Goals, including the eradication of poverty and hunger, quality education, gender equality, decent work, innovation and industry and climate change. SDG 17 recognises the value of multi-stakeholder partnerships for pooling resources, knowledge and expertise to meet shared goals, underscoring the value of universities to serve as fulcrums of development, bringing their capacities for research, innovation, training and community engagement to activate wider stakeholder partnerships. As all the CARPs gained traction and built platforms of engagement bringing together multi-stakeholders, the power of the approach was ignited. In addition to working with TVETs, each of the projects engaged various stakeholders from agencies including national and local governments, extension and veterinary services, NGOs, community-based organisations, civil society, women's groups, farmers' groups and the private sector. These relationships were different for each CARP, depending on the local needs and context. In the Uganda rice CARP, for instance, strong collaboration developed between the National Agricultural Research Organisation, which trained farmer groups in seed production, also with local rice millers, as described in Chapter 2. In the Uganda potato CARP, a local church offered land for field trials and helped to mobilise community involvement. The Zimbabwe CARP and the Ghana pineapple CARP both worked closely with extension agencies. The baobab CARP in Benin involved NGOs and the national nutrition agency. The Namibia bush value chain CARP also collaborated closely with NGOs and with the meat industry. In Sudan, the beekeeping CARP mobilised relationships with Sudan's Agricultural Research Corporation, the United Nations Food and Agricultural Organisation and other national and international agencies. For all the projects, relationship building with wider stakeholders has helped to sustain the momentum of collaborations and activities, contributing towards the longer-term sustainability of the value chain innovation platforms.

In the Botswana safflower CARP, farmers' groups organised themselves to influence government policy and lobby for investment in seed oil production, as covered in Chapter 2. While the university was the catalyst for the safflower value chain development, the farmers are now driving the process. The socio-economic impact of producing animal feed from safflower has also been strong, relieving the burden of costly imports. As a drought-tolerant crop that grows in saline soils and produces a multitude of different applications from fresh greens to potentially high-value speciality products such as health supplements, the diversity of uses and potential for further growth is encouraging. CARP Principal Investigator Professor Vallantino Emongor (2022) commented:

> The university has become more visible, farmers are flocking to the university, I am getting constant phone calls and my office is becoming as crowded as a clinic. We have produced all the information to give farmers in electronic or hard copy brochures, and the students are assisting with this.

In the Kenya potato CARP, close collaboration with the Nakuru County local government helped to build greater coherence in the potato sector. Before the project began, potatoes were not on the local government's radar, even though they are one of the county's most widely grown crops. This changed through the efforts of the CARP. In 2013, as many functions of Kenya's national government were devolved to the local level, the Ministry of Agriculture in Nakuru County had to step up to play a much larger role in the governance of the agricultural sector. The CARP has been running alongside a parallel process of building local government capacity to fulfil its new mandate.

In 2018, the county convened a Nakuru stakeholders forum at Egerton University and also signed a Memorandum of Understanding with the National Potato Council of Kenya – an organisation dealing with policies and strategies concerning the potato sector, including market access for farmers. New national regulations governing the Irish Potato sector were then unveiled in May 2019, which prompted the Nakuru County government to adopt its own county-wide policy. The CARP has played an important role in these processes, providing essential feedback towards the development of both of these policies. The Nakuru Potato Strategy, adopted in 2019, is a prime example of how multi-stakeholder collaboration through the CARP translates into practical working relationships, both developed through and enabling of productive work on the ground that has many levels of impact. 'When it comes to seed potato production, Egerton University is a reference point for us in terms of technical skills, capacity building and research. We work hand in hand with them', remarked one Nakuru County Government official (Lindow, 2020).

In Ghana, the pineapple CARP harnessed technologies including drones and mobile phone platforms to help small-scale farmers build their technical and production capacities and link with agro-processors, commercial producers and input suppliers, as well as government extension services. The CARP aimed to stimulate a local value chain for smallholder pineapple production. Ghana was once a significant exporter of pineapple but lost its market in the late 1990s as Costa Rica gained dominance with the new MD2 'sweet gold' variety combining sweetness with better consistency and a longer shelf life. In recent years, the Ghana government saw new potential for a pineapple industry to supply local demand for juice and other products, but local farmers had abandoned production and moved on to other crops. Through mobilising a multi-stakeholder platform, the Ghana CARP has sought to revive local production and stimulate new industries along the value chain to make pineapple production viable for smallholder farmers. Through the CARP, a pineapple agro-processing hub has been set up on the University of Cape Coast campus to incubate new pineapple enterprises, including juice production, and strengthen relationships with farmers and the private sector into the future.

With guidance and support from Principal Investigator Prof. Festus Annor-Frempong, Shaibo Zikiru, a PhD student in the CARP, used several technologies to help integrate different stakeholders in the platform. Pineapples are heavy feeders requiring copious nutrients from the soil. Flying a drone equipped with sensors and spraying capabilities over pineapple fields, Zikiru could determine and deliver the specific nutrient requirements of each plant, optimising the use of costly inputs. Using an innovative combination of computer assisted personal interviews (CAPI) with focus group discussions, interviews, participatory observation and value chain analysis, he collected data from different pineapple producers, processors, input suppliers, marketers, consumers, municipal and district officials which helped to assemble a rich and comprehensive picture of opportunities and needs from different perspectives. The mobile phone platforms then connected farmers with different private sector and government actors for more efficient exchange of information as well as technical advisory support. Using these technologies to support farmers and share information across the value chain with multi-sector stakeholders, Zikiru was able to develop a niche as a researcher enmeshed in the emerging, growing, participatory dynamics of the value chain. Now employed as a researcher at the University of Cape Coast in Ghana, he is establishing a business in the Department of Agricultural Economics and Extension to provide advisory services using drones and CAPI for digital data collection. The usefulness of drone services to optimise nutrient delivery has become increasingly apparent with the ongoing Russia-Ukraine war, which has contributed to inflation in Ghana, raising the cost and limiting the availability of imported fertilisers and other agricultural inputs.

In the South Africa wool CARP at the University of the Free State, meanwhile, efforts to strengthen the participation of smallholder farmers in the sheep wool value chain brought together diverse science and innovation capacities from the University of the Free State to help small farmers compete in the global marketplace and develop spinoff industries from their lower grade wool. As in other value chains focused on by CARP projects, small-scale wool farmers were being fleeced by buyers who bought their small quantities of ungraded, unsorted wool for a fraction of the potential value. The multi-sector partnerships with the National Wool Growers Association and wool cooperatives helped university researchers to train small-scale farmers in shearing, sorting, classing and packaging the wool by quality in order to reach export markets directly. From the leftover lower-quality wool which could not be sold to lucrative export markets, spinoff enterprises were developed through which local women have been trained in additional crafting skills, such as carding, spinning, weaving, dyeing, sewing and knitting (RUFORUM, n.d.).

Establishing a small wool processing factory on the university's experimental farm, the project provided comprehensive training and skills development in shearing, classing, dying, washing, spinning and weaving wool. The project was strengthened by partnerships with the provincial Departments of Agriculture for the Free State and Western Cape provinces, the National Wool Growers Association and two large wool cooperatives. The momentum and visibility of the projects attract partners to the university. At the same time, as the university leadership perceives success, the lessons are used to influence the wider university culture. As the university's Rector and Vice-Chancellor Professor Francis William Peterson has written: 'In essence, it is all about linking the best of the research, teaching and learning skills of staff and students to specific learning and development needs of society' (Lindow, 2022).

According to CARP Principal Investigator Professor Jan Swanepoel, multi-sector partners, including a local bank and the wine growing industry, have been attracted to the project recognising its potential for innovative job creation. Of the 74 women trained in these value-addition skills, the university identified seven women who showed commitment and potential and has employed them at the university's experimental farm to produce products such as made-to-order conference bags. The women are in the process of registering as a cooperative and service provider to the university and will continue to receive mentorship under the CARP. In the process, the university has deepened its learning on practical issues of intellectual property and product development, supporting academics and the institution in taking the next steps towards the commercialisation of products. Small challenges arose as the women retained by the university did not have bank accounts, so the structures had to be put in place to enable collaboration, with support from the university financial team and a local bank.

Conclusion

The stories in this chapter show how the processes of community, interdisciplinary and multi-sector engagement are helping universities to change from within and build vibrant relationships with external partners in order to better serve as fulcrums of development in their societies. From one of the earliest CARPs launched at Eldoret University in Kenya, lessons have emerged on how the learnings and relationships generated through the CARPs can be institutionalised. After the CARP came to an end, the university developed a new outreach centre to sustain community engagements. As outlined in Chapter 1, the CARP had focused on strengthening farmer cooperatives to increase their bargaining power and abilities to engage with the university and other stakeholders.

As Professor Abigael Otinga (2021), a CARP researcher who helped to establish the new outreach centre, reflects:

> The outreach centre was born out of a continuous engagement with communities. We realised that project funding comes to an end, and then university researchers close up shop and return to their offices. But the farmers still have to produce, find markets and continue their dialogues with government agencies. Through the centre, the structures and policies are in place to ensure the university has continuous engagement with communities. Whether or not there are project funds available, the engagement has to continue. Without engagement, we will be leaving the farmers where we found them. Many relationships have been strengthened through the CARP, and we have learned that we cannot do things in isolation. There is no way we can work as university researchers without involving the county government. They are in a greater position to push the agenda at a county level. The centre is trying to keep all these relationships together. We have a small budget to work with the county extension officers, and also to link the students to communities so that it is a win-win for all.

Through the CARPs, universities have been steadily and incrementally learning to reform systems and policies, inculcate an entrepreneurial mindset, reform the curricula and create space for communities within the institutions. Research has been re-oriented to support the growth and relevance of universities as fulcrums of development. They have demonstrated thought leadership and contributed to new government policies and industrialisation processes. In the process of these changes, universities are coming to understand success in a new light. Through the Uganda pig CARP, for example, farmers have seen that locally adapted solutions can be just as effective as imported solutions which are often inappropriate for rural communities because of costs and infrastructural constraints. In such cases, the CARP approach has begun to transform the learning environment. Building on the classical ideas of participatory action research they are becoming a fulcrum for stakeholders. Universities are bringing a focus on entrepreneurship and leadership to address the African context where youth will need to be the ones who drive transformation and growth, equipped with a breadth of leadership and soft skills outside the ambit of a traditional, discipline-based university education. The next phase of evolution in the CARPs is how they are now informing the transitions from projects to helping African universities design long-term strategic research programs for Africa's development, as discussed further in the Afterword.

Bibliography

Elhag, M. 2021. Personal Interview.

Elhag, M. 2022. Personal Interview.

Emongor, V. 2022. Personal Interview.

Kraybill, D., Lynam, J., and Adipala, E. 2021. *Transforming Tertiary Agricultural Education in Africa*. CABI, Oxfordshire.

Lindow, M. 2020. 'TAGDev Stories'. PDF. RUFORUM. https://www.ruforum.org/sites/default/files/RUFORUM%20TAGDev%20Stories.pdf.

Lindow, M. 2022. 'The RUFORUM Journey'. RUFORUM. PDF. http://repository.ruforum.org/system/tdf/The%20Ruforum%20Journey.pdf?file=1&type=node&id=40580&force=.

Mandumbu, R. 2022. Personal Interview.

Mugisha, J. 2021. Personal Interview.

Mugisha, J. 2022. Personal Interview.

Otinga, A. 2021. Personal Interview.

RUFORUM. 2021. 'Gulu Pig CARP'. Video. https://ruforum.org/impact/project/gulu-pig-carp-2021/.

RUFORUM. n.d. 'University of the Free State Wool Value Chain'. Video. https://ruforum.org/impact/project/gulu-pig-carp-2021/.

4

LEARNING IN MOTION

Megan Lindow, Anthony Egeru and Kay Muir Leresche

Introduction

The CARP approach naturally generates innovations and adaptations through its flexible, iterative and emergent structure. Communities and researchers have amassed rich learnings from building and developing value chains together. The impacts of this work are diverse and multi-faceted. Learning was a critical component of the platforms for students, researchers, farmers and other stakeholders. For students, the experience provided invaluable experiential learning through doing, enriching both their studies and their personal lives. It helped them develop skills and networks and develop business and research opportunities, inculcating a mentality of service and strengthening their personal growth and employability. For Principal Investigators and researchers, shepherding the learning platforms meant stepping into rigorous and demanding roles they were never trained for. They had to invest substantial time and effort in building relationships, nurturing partnerships and balancing the diverse roles and responsibilities of different stakeholders and enable the growth and development of farmers' self-determined activities within value chains. Engaging in these complex, multi-stakeholder processes through the CARPs led researchers and students to see their roles differently. These engagements have begun to demonstrate the value of action-oriented multi-stakeholder collaborations to participating universities, students, farmers, local governments, NGOs and industries.

Some of these early lessons were articulated by Pasape (2022) reflecting on the previous dairy CARP in Tanzania:

> The engagement of various actors was intended to create a platform for scaling up innovation and provoking meaningful changes in livelihoods, organizations,

DOI: 10.4324/9781003387497-6

business and market relations, and policy configurations that impact dairy production at regional and/or national levels. The CARP recognizes that partnerships are essential for scaling up innovations and research; and as a university, acting alone in the design and implementation of scaling up innovations is not effective in sustaining engagement with communities.

A number of the CARPs became anchors of practical training for their universities. As practical lessons emerged from the CARPs, these were gradually absorbed into curricula, making classroom learning more current, relevant and connected to real issues. In the Free State in South Africa, for example, faculties as diverse as economics and occupational therapy took inspiration from the CARP to generate their own project-based learning. Economics students began writing business plans based on the real experiences provided by the CARP.

For farmers and communities, the CARPs also introduced valuable learning opportunities. Because the learning was continuous, participatory and inclusive, there was never a need for researchers to 'return' to communities to disseminate their findings. The farmers had participated in producing these findings and were already busy integrating them into their next initiatives. As several of the CARPs worked to introduce new crop varieties and other techniques unfamiliar to farmers, this participatory learning and training in communities was essential. In the Egerton University Kenya cassava CARP, for example, farmers accustomed to growing maize had to learn new techniques and practices for growing cassava. They had to learn to manage two crop diseases they were unfamiliar with: cassava mosaic virus and cassava brown streak disease. They needed holistic support to learn about growing cassava and to access the necessary resources for growing it sustainably. For example, many farmers did not understand the issues of toxicity that sometimes arise when consuming cassava. In this sense, field trials with farmers became learning laboratories where all participants absorbed important knowledge.

With emphasis on relationship building and social learning, the CARPs demonstrated that the university's catalytic role is not only about co-researching locally adapted science and technology innovations with communities, but also is about holding and activating the space among a variety of stakeholders to move together towards common objectives through which valuable scientific knowledge is continuously and iteratively produced. As such, the learning becomes a feedback cycle that continues to produce value for communities, universities and other stakeholders through the generation and uptake of innovations and new knowledge. The multi-stakeholder character of the CARPs offered a critical learning space to meet the depth and interconnectivity of complex challenges such as climate change, youth unemployment and a fourth industrial revolution. They also provided valuable connections for students to build networks and to understand how communities, different government agencies, civil associations and private enterprises operate.

Through the CARPs, learning together involved multiple aspects of building relationships and social cohesion, managing diverse teams of actors, developing locally appropriate innovations and adaptations in response to emergent challenges and empowering students as lifelong learners, changemakers and entrepreneurs. The experiences of the CARPs detailed in this chapter highlight another aspect of the universities' roles as fulcrums for development, focusing on the importance of social learning among diverse actors with different perspectives on systems for developing knowledge, resilience and other capacities to respond to complex challenges and promote sustainable development.

Active, experiential and multi-dimensional learning

The magnitude of global and local challenges that the CARPs engaged with raised the need for active, experiential and practical hands-on learning to drive complex projects forward. Most communities that participated in the CARPs faced deep and interconnected challenges of climate and environment, food security and livelihoods, youth unemployment, and access to markets and capacity. To contribute effectively under such circumstances, students and researchers needed to develop a full range of skills, a collaborative mindset and sensitivity to the local context. In short, the CARPs provided a ripe environment for experiential learning premised on the idea that people learn in more depth from their direct experiences and activities than they do from theory-based learning alone. Direct experience provides a catalyst for deep learning through a cycle of self-observation and reflection, sensemaking and idea formulation followed by active experimentation to test new ideas (Kolb and Kolb, 2005). Through their research, students gained a depth and breadth of experience and skills in leadership, fieldwork, community engagement, facilitation, problem-solving, project-based learning and critical thinking as well as in practical and technical areas. Their roles in participatory action research processes also required that they develop self-awareness, communication skills and emotional intelligence.

BOX 4.1 STUDENT EXPERIENCES IN THE UGANDA RICE CARP

For Irine Akite, a PhD student in the Uganda rice CARP, the transdisciplinary multi-stakeholder platform provided supportive structures for learning and enabled her own research to have a wider impact. She comments:

> In the CARP, I had attention from the Principal Investigator and was also mentored by several others in the platform. I had the experience of building a relationship with the community. I had to define problems with the community. I had to deliver something that would change their lives. The CARP built my capacity to generate interventions with the community and to work as part of a team of project implementers. I also mentored master's students in the CARP. My research looked at product upgrading and also looked at the upgrading of relationships in the rice value chain. From what I observed, individual farmers are too small to influence anything. They can't influence

the market, they can't influence quality, and their bargaining power is low. They need strong relationships with other farmers, but they also need strong relationships with other value chain actors such as traders and processors. These relationships can support them in areas where their relationships with other farmers cannot, such as in accessing innovative finance.

(Akite, 2022)

Her colleague and fellow PhD student in the rice CARP, Irene Akidi[1] (2022), reflected similarly on the processes of experiential learning:

As I kept interacting with farmers, the ideas that I had in my head kept changing. My ability to adapt and solve the needs that I actually met with on the ground was what enabled me to make a difference. When farmers see you making a tangible difference, they learn to share knowledge with you. They come to trust that working with you will be easy. In the rice CARP it was important to start with setting expectations but also letting people see for themselves. We engaged the farmers, the TVET and government agencies from the beginning and agreed on the roles and responsibilities of everyone involved in the project. Each farmer group came to us with their needs. We had a participatory evaluation of the rice varieties in the field with farmers, students and researchers. We helped one group draft proposals. We helped another group develop a local seed growing business. We had one group that bought a tractor to handle the expanded acreage that they had. I believe that as students we played a very big part in the CARP. Now we are parboiling and packaging the rice.

We have had challenges, but we have also developed the ability to adapt and address the challenges together with farmers. For me, it has been a great experience to work under the CARP and see students and farmers collaborating and sharing knowledge.

We attached a master's student to each objective within the CARP. The master's student would act as a mentor to the undergraduate student in the platform. Then the PhD students would act as mentors to the master's students, and the Principal Investigator and the other supervisors would act as mentors to us. So you were never working alone on the project. All the students had to share their ideas, so there was nobody sitting in the corner doing their own thing. All the students learned from each other, and the ability to bring the students together was important. Some students worked across several farmer groups, while other students concentrated only on working with one farmer group. There was a lot of cross learning in the process.

The transdisciplinary aspect of the research introduced richness and variety to the learning, and often brought in new perspectives that would have been missed by others focused only on their small piece of the overall research puzzle. For example, another master's student, Joseph Odhiambo, working on gender issues in the Egerton University, Kenya cassava CARP reflected on his experience (see Box 4.2).

BOX 4.2 A CARP STUDENT BRINGS IN A GENDER PERSPECTIVE

My master's thesis looks at the influence of gender disparities on participation in the cassava value chain. I came up with this topic because men are the custodians of resources in the community. In cassava farming, when the government and development agencies come with programmes that target farmers, they assume that men are the farmers. Even the extension services target men, thinking that the information is going to trickle down to all household members. But women are the ones who work on the farm.

I want to specialise in gender, youth and agricultural value chain development. I want to look at how gender influences access to land, because land is key to agricultural production; but also how it influences access to markets, extension services and credit. After identifying these constraints and disparities, I will be able to recommend interventions to empower men or women, in cassava or any other value chain.

Youth are often looking for income, and women are often looking to feed their families. If we start with women and the family gets well nourished, we can have healthy families. When there is food insecurity in the household, their thinking will always revolve around how to get food. They can't advance beyond that unless they are empowered and become food secure. I think that is where most development agencies have failed...

But gender is difficult to understand – most people think gender is only women, it's all about roles and responsibilities assigned to men, women, girls, and boys in society ... For example, cassava is a traditional crop, which is normally produced for household consumption and left to the women to manage and make decisions. But as cassava becomes a cash crop, men come in and start domineering. With the value chain development issues coming in, you see more men starting to make the decisions.

(Lindow, 2020)

Professor Anthony Kibe (2022), Principal Investigator of the Kenya potato CARP, reflected also on the richness of the transdisciplinary learning that brought out new perspectives to better inform researchers' understandings of how systems worked:

> Six of our students were in social sciences, and I found that the information was so important. I found that I need to know household income levels. For a farmer to succeed in the seed business they must have an annual income of not less than [US]$6,000. [*Eds: to be in a position to defer using the income generated in order to invest in basic infrastructure and inputs.*] Subsistence farmers require new materials with each new growing season, but farmers with better incomes are able to continue multiplying their harvests. Those students' research produced good insights that were able to be taken up and applied in a wider platform. That is the importance of sharing learning on the platform.

Learning in the field

The CARP is reflective learning in practice. It provides opportunities not only for learning in the field but, importantly, also gives the students time to reflect as they are analysing and writing up their research. It provides farmers, students, advisors and other stakeholders in the value chain the opportunity to reflect together and learn from each other. This strengthens and expands the learning and helps to build trust while at the same time improving university curricula and learning approaches. The experiential learning in practice creates a better understanding of research, innovation and intervention needs, and it helps the service providers deliver more tailored solutions creating a win-win scenario for all. It also provides windows into future opportunities for research and enterprise development.

The activities of the University of Nairobi cassava CARP took place along the eastern Kenyan coast where cassava is a staple food, yet its powerful potential as a key crop for food security, poverty reduction and climate change adaptation had yet to be appreciated. Cassava is hardy and adaptable to erratic and variable growing conditions. Its high productivity throughout the year, relative tolerance of poor soils and drought and the diversity of subsistence and commercial applications make it an important crop for smallholder farmers reliant on rain-fed agriculture. The CARP interventions took shape from an awareness of how continuous recycling and exchange of seeds and planting materials among farmers was contributing to the spread of cassava viral diseases and bacterial blight. The CARP took the initiative to build a clean seed unit and a tissue culture laboratory at the university for cleaning the cassava germplasm and a system for distributing clean planting materials to farmers. The project focused on field training in micropropagation with women and youth, certification of planting materials and value addition and product development to enhance productivity, incomes, food security and nutrition (RUFORUM, 2020). The learning orientation of the platform was essential for linking all the different threads of relevant knowledge together for greater application and impact. As whiteflies were disease vectors, researchers and students focused on mapping their genetics, monitoring their population dynamics and testing push-pull approaches to pest controls in farmers' fields. In another thread of research, PhD student Florence Munguti focused on developing a field-based diagnostic tool to be used to certify clean cassava planting material and help with the early detection of cassava brown streak virus in farmers' crops so they could intervene before it was too late.

For students, experiential learning in the field brought deeper levels of insight and the ability to draw connections that could only be made through participatory interaction. The Nairobi cassava CARP was also an international one. It worked with another RUFORUM university and extended into areas of eastern Uganda, where plant breeding master's student Faizo Kasule from Makerere University in Uganda conducted field work and reflected on his experience. He wrote:

> As a student pursuing plant breeding, I realised the need for more improved varieties that are not only high yielding and disease resistant but also with 'good' taste attributes ... I also realised that farmers need to be sensitised on the benefits of starting with clean/virus free planting materials obtained from credible sources. Such material could be obtained from tissue culture laboratories and multiplied in isolated places before dissemination of material to farmers. This underpins the importance of community based seed systems development in these rural areas.
>
> *(RUFORUM, 2020)*

The multi-disciplinary environment of the university both held the emerging knowledge and extended the reach of these key insights through reforms to the curricula, teaching and training in collaborations with other stakeholders. In terms of the economics of cassava, for example, efforts to relieve over-reliance on maize in drylands areas are well and good. However, maize is likely to still be more profitable as an established commodity in local food systems until the cassava value chain becomes developed well enough to stimulate an equivalent demand through new markets, linkages and value-added products. Looking to improve profitability, another strand of the CARP developed chicken feed formulations using underutilised cassava peelings in place of maize, which increasingly has to be imported from neighbouring countries to meet demand. In the COVID-19 pandemic, disruptions exposed important weaknesses in Kenya's food system as key foodstuffs requiring transport became unavailable. Because farmers had already been trained in cassava value addition when the pandemic hit, they were able to compensate for a lack of availability of other fresh vegetables by increasing their utilisation of cassava products which remained available and accessible, including the roots, stems, leaves and peels (RUFORUM, n.d.).

All of the knowledge produced through the CARP was available for the farmers, TVETs, women's groups and students to integrate and develop new niche specialisations for themselves and entrepreneurial start-ups to serve communities. Before the CARP, farmers were complaining of limited markets for their cassava, while cassava traders were complaining of shortages of cassava to sell. The learning processes of the CARP helped to establish new connections and generate new opportunities. In one example, a medical student who came from a family of subsistence cassava farmers had struggled to afford his university fees which had forced interruptions to his studies several times. With advice from CARP Principal Investigator Agnes Mwang'ombe, he was able to tap into a much more lucrative opportunity in trading cassava roots, which were in high demand (RUFORUM, 2020). Other young people from communities developed businesses as diagnostic experts helping farmers identify crop diseases in their fields using the diagnostic knowledge and tools developed through the CARP.

In the Kenya Seed Potato CARP, meanwhile, the multi-stakeholder platform involving researchers and students from Egerton University, the TVET Baraka Agricultural College, the National Potato Council of Kenya, the Nakuru County Government Ministry of Agriculture, the state seed potato merchant ADC-Molo and the Nakuru Smallholder Farmers Association worked with six local small scale farmer groups to pool resources and strengthen capacities to supply seed potatoes across the region. The CARP interactions with farmers revealed the intensity of challenges they were facing as they had to make key decisions about when to plant and where to invest their limited resources in the face of mounting uncertainty and risk from climate variability.

As CARP PhD student John Ng'ang'a reflected, unpredictable rainfall patterns were just one of many variables needing to be weighed by farmers. Weather variability was also exacerbating pest and disease outbreaks, which farmers then had to control with costly applications of herbicides and fungicides or else risk losing their entire crop. Such dilemmas were all the more vexing to small farmers, John explained, because the outcomes of their labour were so uncertain. An untimely drought may come along and wipe out a crop that the farmer has sunk all her resources into. A farmer confronted with a sudden infestation of blight or some other pest may be forced into debt – or may simply not have the means – to buy chemicals to spray on the fields.

As researchers were discovering, no amount of hard work or careful planning was ever enough to guarantee farmers a harvest. Even in good seasons, because farmers planted at the same time there would often be a glut in the market and, without proper storage facilities, they would be forced to sell at rock-bottom prices. Such complexities are testament to the intense pressures faced by people and the land. Ideally, land used for potato production should rest between growing seasons. But for a farmer living hand to mouth on a couple of acres of land or less, this was not possible. As the CARP process has grown, the understanding of such challenges, and awareness of the need to find solutions that fit the farmers' specific contexts, has deepened. Through the CARP, some farmers built their own seed stores and explored ways of water harvesting, as well as combined their efforts with neighbours to minimise costs and maximise profits.

At Baraka College, crops tutor Percy Njeri also was working with farmers to help them minimise their use of chemical inputs. Passionate about soil, microbes and traditional cultural practices of farming, Percy experimented with different organic techniques for restoring the soil. 'If your soil is fertile you will not have a lot of problems', she said.

> We often say pests and diseases attack crops, but it's often stemming from a deficiency in the soil. If you feed the soil then the plant will be strong. It will have defences, because you have already given it immunity through the soil. We want chemical use to be the last option. We are trying to promote cultural methods, and biological methods to conserve our natural predators to pests.
>
> *(Njeri, 2019)*

Building relationships and managing teams

The CARP model puts relationships at the centre of learning. The processes of building community engagement and bringing in other partners in local government, NGOs and the private sector produced rich learning environments for students, researchers, institutions and farmers alike. The reorientation of approach from extractive and top-down 'technology transfer' and 'outreach' to collaborative and participatory research with communities imposed serious demands for learning, reskilling and intentional changes to mindsets and institutional cultures. At the same time, the multi-stakeholder platforms allowed for wider and more comprehensive uptake of the learnings that did emerge.

In every CARP each student's research brought out a different agronomic, value chain, marketing or socio-economic aspect. When combined, these different aspects contributed in diverse ways to the overall strengthening of the value chain. For example, in Ghana, one student discovered that harvesting pineapples in the morning for juicing improves both the taste and nutrition levels. This knowledge then could be taken up commercially by students and extension workers working with farmers.

As noted in Chapter 2, several of the CARPs built on pre-existing government initiatives, including the Uganda rice CARP and the Kenya cassava CARPs. These initiatives signalled from the inception the willingness of researchers to work within wider multi-stakeholder collaborations in order to amplify the impacts of their projects. This approach recognised the power of transdisciplinary research as a knowledge-producing, social learning system. In the Uganda rice CARP, for example, participation of the national research organisation was mutually beneficial, as they provided the different rice varieties for testing in field trials, and at the same time benefited from the farmers' insights as to which varieties they prefer and grow best in their fields. Two farmers' groups decided later to enter the seed production business and the NARO was able to train them to a level where they could obtain recognition for their quality seed.

Overseeing the projects and brokering all the different relationships required that the Principal Investigators become jacks-of-all-trades. Any given week might involve organising meetings, waiting patiently in government offices, scrambling to the field for planting days with farmers after the rains, organising logistics and keeping track of expenditures, supervising students and weaving together the different multidisciplinary research threads of their work. Most of the Principal Investigators and associated researchers on projects came from hard sciences backgrounds and had to rapidly upskill themselves in meeting all these demands. As one Principal Investigator quipped, 'I'm becoming more of a community organiser than an applied scientist' (Lindow, 2020). Establishing the platforms placed heavy demands on their time and energy and required them to attend to intricate processes of relationship building. In the Namibia CARP, researchers spent time building relationships with the nearby TVET principal, only to have him leave his position. After all the time invested in building the relationship to involve the TVET in the project, the conversation had to start anew with his successor. A number of factors hampered the new relationship, including COVID-19 and logistical issues such as the distances between institutions, which meant that although there was some collaboration the institutions never succeeded in establishing the deep collaboration initially envisaged. Such were the facilitation and transaction 'costs' of establishing the platforms and brokering the necessary relationships to build momentum and sustain these activities.

Learning as an iterative cycle

The CARP approach of one thing leading to another (see Chapter 2) reinforced a cycle of continuous learning across multiple levels. Often as new challenges arose in the platforms, they became opportunities to reflect, challenge assumptions, do things differently and try new approaches. In some instances, challenges were met with creative, adaptive or innovative solutions. In the Uganda rice CARP, for example, a season of heavy rainfall led to the rapid growth of vegetation in pastures where livestock browsed. With their normal pastures overgrown, the animals of local farmers suffered from food shortages. From this problem, one farmer group spotted an opportunity to produce animal feed from their fermented rice straw. As each of the different farmer groups chose different areas of the value chain to focus on – from parboiled rice to rice seed to charcoal briquettes and animal feeds produced from rice straw – new lessons emerged from their experiences which could be shared across the wider platform.

In this way, the CARPs became both experimental platforms and scaling laboratories, where new ideas emerged and could be tested, scaled out and inculcated into the learnings and operations of institutional partners. These were transdisciplinary spaces for testing, piloting, incubating and gathering knowledge to help the uptake of new ideas and innovations. Sometimes these new ideas emerged from the barriers and limitations of local contexts. Understanding both the particular constraints and untapped resources of local communities further enriched the process. In the Uganda pig CARP, for example, conventional 'first world' solutions to the problem of foul-smelling piggeries were unaffordable and inappropriate for local application. This led to the adoption of the pig houses with deep litter floors doused with the IMO solution adapted from the global scientific literature. Similarly in the production of AI techniques, the constraint of access to costly imported extenders for the pig semen led to experimentation with locally abundant sources of coconut water as a replacement. Where conventional solutions did not fit or were out of reach, university researchers were able to adapt scientific knowledge and innovation to find locally appropriate solutions to challenges (RUFORUM, 2021).

In the pig CARP, combining different innovations produced healthier pigs and better-quality meat, using methods that were within farmers' means. Sows' litters increased in size. Then marketing interventions further enhanced the impact of these innovations by opening up new entrepreneurial avenues in more lucrative value-addition streams for farmers. As farmers learned what the market wanted, they could make better decisions. As Principal Investigator Professor Elly Kurobuza Ndyomugyenyi (2022) remarked:

> The CARP has facilitated a mindset change towards new innovations. People thought exotic technologies were superior to the indigenous. But the mindset changed when locally formulated feeds are performing as well as the costly imported feeds. Farmers can see for themselves. Out of school youth are seeing these developments, and they are working to extend these innovations to the community.

CARP activities also helped to build social cohesion in the communities. For example, the pig houses with the IMO solution in the deep litter floor were labour-intensive constructions. Building these houses together, local pig farmers demonstrated their willingness to work together and learn together to accomplish what none of them could do on their own.

BOX 4.3 A YOUNG LECTURER'S PERSPECTIVE ON KNOWLEDGE CO-CREATION WITH COMMUNITIES

Daniel Okello, assistant lecturer at Gulu University, was involved in the pig CARP, and reflected:

> Working with farmers I realised they are often faced with all these issues – the health challenges, zoonotic diseases like tapeworms needing controls, and marketing challenges. The farmer knows he has to invest in biosecurity, which he can't afford, so he looks for an alternative that is more applicable. You go with your research of marketing and they start to bring new issues of feeding or health management which probably wouldn't have been the

focus of your study. Now you have to widen your research area, or try to be innovative – because one area affects the other. They can't sell their pigs profitably if they aren't producing in the best way possible. Participatory research in that case becomes very critical – it helps address the challenges at farm level and also in generating knowledge at farm level and at the same time disseminating the research to the same farmers because it is also context specific...

During the engagement we were able to come up with feed formulations from local free resources that are available within the local communities that farmers can easily access. For instance, farmers didn't know they could mix sweet potato vines with other ingredients to feed their pigs, but through research and through the participatory approach they're able to understand that this resource that is available can actually be utilised properly. They see how it works compared to other findings. I think it gives farmers a positive mindset, they appreciate the whole process, because you find they have been involved in the generation of knowledge. At the end of the day you don't need to go back to disseminate the same knowledge, because they have it, you did it together with them. The only dissemination now would be to take it to another community that was not part of it. Through examples we're able to demonstrate that it works. Others have adopted better feeding practices and are able to get more money.

(RUFORUM, 2022)

In these processes of experimentation and learning, the failures and setbacks could be just as instructive as the successes. In some cases, pursuing certain goals and relationships revealed practical constraints and limitations. The Kenya potato CARP, for example, had tried to organise farmers to obtain official certification for their seed from the national regulator KEPHIS. In practice, however, the requirements of this process were too stringent and expensive for the farmers. Through the exercise, they learned that it made more sense for them to focus on producing clean, high-quality seed without certification for their own use and to sell to other local smallholder farmers.

Other CARPs meanwhile showed disappointing results from their efforts to engage with the private sector. In the Egerton University, Kenya cassava CARP, for example, overtures to East African Breweries to purchase farmers' cassava and introduce a line of beer produced from cassava stalled. Initially, the idea raised excitement from the company, but expectations were unrealistic. When it became clear that it would take more than two years for farmers to reach the capacity to produce commercial volumes of cassava, the company's interest waned. The experience highlighted the wide chasm in expectations and capacities between the commercial and smallholder sectors, which is widening further as trends of urbanisation and dietary changes reshape food systems and introduce greater demands for modern, processed and convenience foods. Yet these 'failures' and disappointments also helped researchers learn and fine-tune where best to focus their efforts.

CARP Principal Investigator Richard Mulwa (2022) commented:

> When we started this project, we had a cassava cooperative that was moribund in Subukia. We realised the cooperative had been bought machinery by the World Bank which had never been operative – we facilitated the connection of electricity. In addition, farmers are already hand chipping and drying at home and so there is potential for mechanical chippers on motorbikes which can be a business for youth doing the rounds among farmers.

Through the CARP, farmers have identified a cassava variety that has good commercial potential for industrial use because of its high starch content. Such simple learnings that continuously emerge from the multi-stakeholder platforms generate new threads for research and development to explore. In this way, the collaborative, cross-disciplinary aspect of the CARPs was essential to the learning and discovery that enabled farmers to move and grow. With the wide, cross-cutting and iterative focus of the CARPs, continuous learning often served to make promising new avenues and next steps for the research clearer, as these next steps were informed by the knowledge production thus far. The flows from knowledge to research to business were not always seamless, but even the challenges and obstacles generated new learning along the way. This was the power of iteration in a wider system that allowed learnings to be absorbed and taken up and built upon further.

In the bush value chain CARP in Namibia, learning similarly emerged in non-linear ways. Namibia is a major beef exporter, but problems of land degradation known as bush encroachment are threatening grazing lands. Many factors contribute to bush encroachment, including climate change and poor land management. Initially, the CARP was intended to establish bush clearing to restore grazing lands, but it shifted focus from clearing to developing value addition from the bushes and trees contributing to the encroachment problems. Changing the orientation of the project opened the way for innovative product development. Researchers worked on formulating animal feed from the cleared branches and could also incorporate another student who developed mushroom cultivation techniques using the larger logs as a growing medium. Namibia CARP Principal Investigator Professor Simon Angombe (2022) remarked:

> The game changer for the community is the feed formulations. They have seen that they can use bush-based feed for their animals and don't have to buy commercial feeds. Now we need to find other communities that are vulnerable to climate change and train them and scale up training.

Responding to future needs

Participatory research helped to shine a light on previously ignored or neglected research areas of importance to communities, raising institutional awareness of these issues. In Zimbabwe, the CARP's focus on small grain crops raised awareness of neglected research areas that universities are uniquely positioned to work on. Processing of the small, fine grains is labour intensive, partly because research systems have not focused on developing the value chain and infrastructure to facilitate processing in bulk for the widespread uptake of these food sources among the population. Globally and locally, research systems have focused on maize and a limited selection of other staple crops for Africa. However, issues of increased drought and climate variability as well as health and nutrition concerns raised by an over-reliance on maize in peoples' diets make it necessary to develop alternative value chains. Before maize gained dominance in local food production systems, people were already working with crop systems and managing soil fertility in climate-resilient ways (McMullin et al., 2020).

Zimbabwe CARP Principal Investigator Ronald Mandumbu (2022) remarks:

> The small grains and legumes are some of the best crops to help communities build resilience to climate change. In my view, these are orphan crops. They were excluded from the design of most processing equipment, but this has the potential to be developed. Through the CARP project we can focus on developing the infrastructure to enable the uptake of these crops to make them a viable food source for a lot of people. When we work with communities, we are promoting the uptake of these products and the community is accepting them. These foods can provide micronutrients and address hidden hunger and issues like high blood pressure, because the carbohydrates are released more slowly into the blood. Sesame for example has a very high content and quality of oil. But it became ignored because research did not focus on promoting it or producing the seed. Going back to these crops is building diversity and resilience.

This comment reflects how working with communities to develop and combine the strengths of indigenous knowledge and science and innovation is an important way to build capacities for resilience and adaptation to respond to climate change and other emerging challenges. Mandumbu (2022) remarked:

> Once you are collaborating and sharing knowledge as communities and as the university you are already in a stronger position to deal with the shocks that are coming. Even in the most intense drought you will continue to make the *Zizyphus* porridge because the ingredients are all climate resilient. We don't have a single ingredient that requires wet growing conditions. Some of the ingredients are from trees which have deeper root systems. They are water efficient and

> fertiliser efficient. We also use local manure for the crops – treated to make sure the seeds within it are dead.

The next learning curve for Zimbabwe and other CARPs is to fully commercialise the products that have been developed and tested. Product development has generated enthusiasm in the communities. Yet the capacity for full commercialisation within the university remains limited. As Mandumbu (2022) remarks:

> We cannot do full scale production while we are still roasting using a pot. We have developed other methods. Our porridge and masau drinks are ready to hand over to the private sector, but reaching an agreement is a challenge. There is a lot of mistrust of the private sector. Communities have gone through a lot. They have grown crops people say they will buy and then they disappear. Trust is a major challenge.

Undergoing such processes of rich experiential learning, universities have begun to adopt new approaches to learning in the classroom, engagement with communities and their collaborations in wider innovation systems.

BOX 4.4 A PRINCIPAL INVESTIGATOR'S PERSPECTIVE ON HOW CARPS BUILD INSTITUTIONAL RELEVANCE

As Mandumbu (2022) reflects:

> The CARP project has helped us to be relevant to our community and to the nation at large, and even to our students. Sometimes lecturers haven't updated their lecture material, but things on the ground have changed. What we have learned through doing the CARP has been able to feed into changing and updating the curriculum. Now seeing the advantages of tied-contour water harvesting, sustainable crop production and sustainable use of natural resources – all of those we are now using the results from the CARP to teach our students. It's more practical and relevant to students. So we have a better informed lecturer and a better informed student. Then their degree becomes more valuable and more relevant.
>
> Before the CARP we couldn't really teach about innovation, but from the project learnings we can now talk about it with confidence. We can say these are the products we have developed in collaboration with students. We are more relevant as an institution.

> Through the CARP the university has developed products and prototypes with big potential to be commercialised. It has helped lecturers stay up to date and relevant. It has lightened the burden of unemployed youth on the community. The number is low but if two students are trained from the community, it demonstrates to the youth that education can help them. They can be role models for other young people. It has all come from the CARP.

Conclusion

The platform approach has enabled learning to surface and be utilised within broader systems of collaboration. Through engaged and experiential learning, students gained a much richer and fuller breadth of experience and skills, which many have successfully applied to new business ventures that serve communities. Universities formed key relationships that were sustained by social learning processes with a multitude of different partners. Researchers learned to be jacks-of-all-trades, no longer confined to their disciplinary siloes, but able to engage with a broad spectrum of local issues and challenges. These processes held broader relevance in terms of building the universities' capacities to respond to complex societal challenges including climate change and nutrition security. Engaged in these multidimensional processes of learning, universities were able to advance their own capacities to serve as fulcrums of development. While the CARPs have shifted the emphasis from traditional research to a practical focus on real-world change and development, the quality of scientific outcomes has not suffered, and the journal publications and research outcomes have been robust as can be seen in Part II, see particularly Chapters 7, 8, 10 and 21.

Note

1 Irene Akidi is a research assistant on the rice CARP but is enrolled for her PhD at Egerton University in Kenya.

Bibliography

Akidi, I. 2022. Personal Interview.

Akite, I. 2022. Personal Interview.

Angombe, S. 2022. Personal Interview.

Kibe, A. 2022. Personal Interview.

Kolb, Alice Y. and David A. Kolb. 2005. 'Learning Styles and Learning Spaces: Enhancing Experiential Learning in Higher Education'. *Academy of Management Learning & Education*, 4(2): 193–212.

Lindow, M., 2020. 'TAGDev Stories'. PDF. RUFORUM. https://www.ruforum.org/sites/default/files/RUFORUM%20TAGDev%20Stories.pdf.
Mandumbu, R. 2022. Personal Interview.
McMullin, S., Stadlmayr, B., Mausch, K., Revoredo-Giha, C., Burnett, F., Guarino, L., Brouwer, I. D., Jamnadass, R., Graudal, L., Powell, W., and Dawson, I. K. March 2021. 'Determining Appropriate Interventions to Mainstream Nutritious Orphan Crops into African Food Systems'. *Global Food Security*, 28: 100465. https://doi.org/10.1016/j.gfs.2020.100465.
Mulwa, R. 2022. Personal Interview.
Ndyomugyenyi, E. 2022. Personal Interview.
Njeri, P. 2019. Personal Interview.
Pasape, L. 2022. 'A Review of Success Factors Behind Community Action Research Program (CARP): A Case of Experiences from Smallholder Dairy Farmers of Lushoto in Tanzania'. *Journal of Participatory Research Methods*, 3(2). https://doi.org/10.35844/001c.37544.
RUFORUM. 2020. 'Voices from the Field: Transforming Cassava Production through Delivery of Clean Planting Materials and Business Development in Coastal Kenya'. Vol. 5. RUFORUM. https://ruforum.org/sites/default/files/VFF%20Vol%20%205%20Web.pdf.
RUFORUM. 2021. 'Gulu Pig CARP 2021, Uganda'. Video. https://ruforum.org/impact/project/gulu-pig-carp-2021/.
RUFORUM. 2022. 'The Movement of Changemakers in Africa'. Video. https://www.youtube.com/watch?v=KZqbTMUJX6A.
RUFORUM. n.d. 'University of Nairobi Cassava CARP+'. Video. https://ruforum.org/impact/project/university-of-nairobi-cassava-carp/.
RUFORUM. n.d.a. 'Egerton University Seed Potatoes CARP, Kenya'. Video. https://ruforum.org/impact/project/egerton-university-seed-potatoes-carp/.

5

CHANGE AGENTS, RIPPLE EFFECTS, WIDER IMPACTS

Megan Lindow, Anthony Egeru and Kay Muir Leresche

Introduction

As highlighted in previous chapters, the participatory, collaborative, interdisciplinary and multi-stakeholder approach of the Community Action Research Projects (CARPs) galvanised universities, communities and other partners to build their capacities to engage Africa's (and the world's) most pressing challenges as outlined in the Sustainable Development Goals (SDGs). The richness, diversity and depth represented across all the CARP platforms, drawn from many different countries, contexts and value chains, highlighted the wide variety of potential applications and spin-offs inherent in the approach. Indeed, as seen in previous chapters and as will be explored further in this chapter, the platform engagements generated learnings and impacts directly relevant to virtually all the 17 SDGs, including most particularly no poverty (SDG 1); zero hunger (SDG 2); quality education (SDG 4); gender equality (SDG 5); decent work and economic growth (SDG 8); industry, innovation and infrastructure (SDG 9); climate action (SDG 13); life on land (SDG 15) and partnerships (SDG 17).

Many practical actions both small and large touched on these various aspects of sustainable development. As students, researchers and other stakeholders embraced new roles as change agents, they continued to discover new ways to build momentum in value chains and contribute towards job creation and household-level food and nutrition security. Universities discovered their potential to drive small-scale agro-industrialisation and train students for employability and relevance while sustaining and upscaling the platforms. As projects gained momentum and showed results, relationships and partnerships across institutions deepened, inviting further collaborations which sometimes resulted in wider policy-level changes.

DOI: 10.4324/9781003387497-7

Communities found themselves strengthened in terms of their livelihoods, nutrition and dietary diversity and their resilience to shocks such as climate change. Through experiential learning in the multi-stakeholder platforms, students gained the skills, networks and confidence to develop their own successful enterprises in baobab, wool products and other niches opened up through the CARP initiatives. Farmer groups succeeded in establishing cooperatives and developing new value-added enterprises and accessing wider resources and knowledge streams made available through the CARPs. In several cases, the CARPs helped to catalyse or contribute to local and national policy changes. These various spinoffs and multipliers illustrate how the particular focus on action research, entrepreneurship, relationships, multi-stakeholder collaboration and leadership helped to create lasting impacts while empowering young people and universities alike as facilitators of development responding to their own particular contexts.

Strengthening food systems, engaging with youth employment, climate change, biodiversity, nutrition security and gender equality

This book has argued for the essential role of universities in meeting complex and interconnected 21st-century challenges. Because universities combine essential functions of training, research, community engagement, innovation and industrialisation, they have a special role in preparing young people for the present and future. As climate change impacts deepen, the capabilities of these institutions to co-explore the dynamics with local communities, understand their impacts in wider stakeholder learning processes, contribute to policy interventions and pass this knowledge on to students will matter increasingly (Fox and Ghandi, 2021; Habiyaremye et al., 2022). Each of the CARPs generated multi-level ripple effects touching on these key areas. As to be further explored in this chapter, a focus in the Benin baobab CARP, for example, on nurturing entrepreneurship in the baobab leaves and pulp value chain catalysed student entrepreneurs committed to tackling Benin's formidable early childhood malnutrition challenges. In the Gulu pig and Nairobi cassava CARPs, dynamic and resourceful farmer-entrepreneurs rose up to stimulate new local industries, generating livelihoods for young people and building community resilience to food insecurity in the pandemic. The greater cohesion and organisation stimulated through the formation and strengthening of farmer groups has the potential to build community resilience to powerful shocks and stressors of climate change, biodiversity loss, flooding, drought and social conflict (Rwelamira, 2015).

Virtually all the CARPs realised important impacts in food and nutrition security, which underpins all other aspects of societal health. This is a particularly crucial area of intervention across sub-Saharan Africa, where one person in four suffers from undernourishment, while at the same time many communities are sitting with highly nutritious yet under-utilised sources of wild, edible foods including baobab which is found across diverse ranges of eastern, western and southern Africa; *Zizyphus mauritiana*, which is also widespread in Africa; the Sudanese famine foods and many others. Through the CARPs, universities have highlighted their relevance through research and innovation that connects all these different needs in communities for nutritious, affordable and accessible foods that also provide local employment and may further enhance food security.

Students as change agents

The CARPs stimulated numerous student-led enterprises, which have had a powerful multiplier effect.[1] Enabling young people to start their own enterprises as they exit the university system goes hand in hand with the universities' emerging roles of driving innovation and industry through the multi-stakeholder platform approach. Providing both inspiration and practical direction, the experiential learning in the platforms opened students' eyes to exciting new opportunities they could develop for themselves.

Marius Affonfere, a PhD graduate of the Benin baobab CARP, combined his passions for food security, community nutrition and social inclusion with an entrepreneurial drive. In his research under Professor Flora Chadaré, he worked with women food processors in northern Benin to develop products to address food insecurity, in particular a high prevalence of iron and micronutrient deficiencies in children under five. He gathered feedback that communities wanted an easily accessible and affordable nutrition supplement and, based on this, developed iron-rich biscuits made from flour fortified with baobab and moringa leaf powder. He reflected:

> During my research the most important thing for me was to ask: what are the most important benefits to the community from doing it this way? That was the question I asked as I designed the proposal and at every stage of the research. What will be important for the community? What does the community need? Knowing what the community wanted, I could focus my resources on that and on enabling the outputs to be used easily by the local community.
>
> *(RUFORUM, 2022)*

Developing the iron-rich biscuits as a business after his graduation, he applied the same principle. Through a US$5,000 small grant secured through the CARP he developed the project and secured further funding from the Kofi Annan Foundation to upscale it. He commented:

> It came from the question – how can we use our local food resources to alleviate the critical issue of iron deficiencies not only in Benin but in Africa? It is very critical. Children are used to eating biscuits, so we can introduce biscuits that are rich in iron and that can help the community. That is the idea behind the business.
>
> *(RUFORUM, 2022)*

The CARPs provided a launching pad for dozens of other students to pursue diverse enterprises that similarly combine a sense of community orientation, entrepreneurial skill and practical know-how to apply new ideas and create jobs for themselves and others. Student entrepreneurial spinoffs from the CARPs include businesses as diverse as seed production in many locations, drone and technology-based advisory services in Ghana, Kenya and South Africa, production of pineapple juice in Ghana, felt jewellery produced from sheep wool offcuts in South Africa, confectionaries from potatoes in Uganda, millet and sesame-based mixtures in Zimbabwe, processed honey in Sudan, value-added indigenous foods in Sudan and mushrooms in Namibia.

Kenneth Kidega, a former master's student in the pig CARP, was motivated to help northern Uganda's rapidly growing population gain access to an affordable protein source and improve the nutrition status of children through the pig value chain. As he explained, the CARP platform innovations helped to produce healthier pigs in a shorter amount of time, making it possible for farmers to improve their incomes dramatically. Through the project, he observed families steadily gaining food security, eating three meals a day instead of two, or two meals a day instead of one. He said:

> A malnourished child will take a longer time to develop mentally, which affects their schooling and their potential for life. If you try to fight malnutrition and let children eat protein, which is needed for brain development, you will develop the nation.
>
> *(Lindow, 2020)*

Even before he graduated, Kenneth had secured land and built his own IMO-strengthened pig houses to begin his own pig production enterprise. With proceeds from his business, he paid school fees for his brother who then joined the business. Kenneth commented:

> The piggery business has made my family more resilient and stable. Both in my community and in the community where the CARP was operating it has had a significant impact. Before the project our survey found that only 20% of people were rearing pigs, but after the project we found that 72% of people are rearing them. Almost all (over 95%) can now sustainably pay their children's school fees. For 70%, the money they have earned from animal rearing has enabled them to expand their crop production. Now local government officials see pig farming in a more positive light and encourage people to rear animals. The farmers are forming a Northern Uganda Pig Farmers Association.
>
> *(RUFORUM, 2022)*

Kenneth has also entered the mobile money business and employs four people.

Through such engagements, students have developed both practical skills and the ability to see processes of capacity development in communities and understand the maturity of these processes. They have identified new opportunities for businesses and support. They have been part of piloting and testing novel approaches and seeing that with the right approach and sensitivity to context, such interventions are transformative both to communities and individual livelihoods, including their own.

BOX 5.1 STUDENT STORY: REFORESTATION IN KENYA

In some cases, the CARPs indirectly inspired students who were not even part of them. Lilian Onyango, a TAGDev master's student in agri-enterprise development at Egerton University, was a product of the university's transformative education approach (see Chapter 6). Engaging with farmers in the field as part of her coursework, she visited a farmer who was part of the potato CARP. The farmer inspired her because he was practising agroforestry. He was maintaining a beautiful green and diverse system of crop production integrated alongside his tending to the natural environment. During the visit, Onyango was struck with a realisation. She recalls: 'People are dying now in Kenya because of drought and deforestation. We asked what we could do as youth, because we are the change leaders of tomorrow, we are not waiting for others to solve our problems' (Onyango, 2023).

The Kenya Government's Vision 2030 strategy calls for a billion trees to be planted to employ young people and preserve the environment. To contribute to this vision, Onyango and several friends established a youth-led NGO called Youth for Green Action Kenya. Working with youth volunteers across the country, they organise clean-ups and reforestation projects, including work to restore ocean and coastal mangrove habitats. She said: 'We want to bring back the green environment we once lived in. We want to eat the wild fruits we grew up eating' (Onyango, 2023).

The organisation's next project will focus on restoring endangered indigenous trees at risk of extinction. And they are gearing towards a mega-project of planting 100,000 trees in Narok County, Kenya with 300 volunteers. The organisation also works closely with farmers to encourage ecological farming practices including agroforestry and promote awareness of how farming can affect larger cycles of water and climate. She explains:

> The rains are not coming and the water levels in the rivers are dropping rapidly. We teach farmers not to plant trees that drain large amounts of water from the soil. We need to harmonise agriculture with the natural environment. So we focus on where to position trees so they are not near a river. Farmers and communities are responding so well to these activities.
>
> (Onyango, 2023)

BOX 5.2 STUDENT STORY: KICKSTARTING A POTATO INDUSTRY IN LIBERIA

Sheku Gbollie was a TAGDev master's student at Egerton University who carried out research in the potato CARP on using cocopeat as a soilless medium for the production of potato minitubers for seed multiplication. Originally from Liberia, he grew up during the civil war and endured a long struggle for education before landing a rare opportunity to do his master's.[2]

Returning to Liberia in 2022 upon completion of his master's, he took the initiative to apply his learnings from the CARP to generating a new potato industry in Liberia, where the potential impacts in job creation, farmer income generation, food security and dietary diversity are high. His work with cocopeat – a growing media produced from grinding and treating coconut husks for the production of horticultural crops – makes sense for Liberia where coconut husks are abundant but not typically utilised, he said.

Potatoes in Liberia are a rare and expensive delicacy; demand is high and supply is low. He said:

> There's a perception that potatoes can't be grown in Liberia. We are focused on rice and cassava production, but we need to bring about food diversification. One of the things people don't realise is that we have highland areas with good growing conditions for potatoes, and also there are potato varieties that will withstand heat, so why don't we try?
>
> (Gbollie, 2023)

Employed in an internship with the Central Agricultural Research Institute in central Liberia, he is working to overcome the administrative and practical hurdles to enable potato production to begin. He is working with the governments of Kenya and Liberia to arrange for the importation of potato seedlings, and trying to source funding for a grinder to enable the production of cocopeat. He said: 'I feel very motivated. This will broaden the application of research in Liberia. It will help people to widen their thinking, and it will bring about job creation and help farmers to generate better incomes' (Gbollie, 2023).

Farmers getting organised

As demonstrated in the lessons from one of the earliest CARPs at the University of Eldoret in Kenya (Chapters 1 and 3), one of the most effective ways for farmers to unlock the power and potential of value chains is to organise themselves, supported by multi-stakeholder relationships in the platforms. In Eldoret as well as in the Botswana safflower CARP, the Kenya potato CARP, both Kenya cassava CARPs, the South Africa wool CARP and the Uganda rice CARP, farmers discovered that by organising themselves into strong associations, they could aggregate their produce and gain power in the market. In the Gulu pig CARP, farmers worked together to build IMO pig houses helping each other gain the benefits of healthier pigs, sanitary conditions and better-quality meat which none could have realised alone. In eastern Kenya women's associations have been active in producing and marketing cassava products as a result of their CARP engagement. In Sudan, the formation of a women's co-operative has contributed to increased incomes and the empowerment of women through improving their household nutrition and earnings. In Kenya, South Africa and Uganda communities built new value-addition centres. Also in Zimbabwe, value-addition centres were built to enable processing and product development with the goal of raising resources to educate local young people in desperate need of skills and employment. In Botswana, farmers organised themselves to lobby the government to invest in the development of a profitable seed oil industry, resulting in new policy initiatives highlighted in Chapter 2.

These measures were enabled by processes of learning, collaborating and building trust among fellow farmers and in wider multi-stakeholder platforms. As farmers accessed training and gained skills, knowledge and exposure to information such as market demand for their products, they were able to take up new opportunities in a coordinated way. Some of these farmer associations existed before the CARPs and were strengthened through the platforms. Others were initiated by the CARPs or were formed by farmers themselves as a result of their participation. Through these strengthened associations, farmers organised themselves to become seed growers, invest in new machinery, venture into new enterprises and scale up their operations. Thus, they began to unlock their powers to stimulate small-scale agro-industrialisation. This had many ripple effects including better incomes to invest in educating their children, creating employment and better positioning themselves within value chains.

In a number of the CARPs, including the Uganda pig and rice CARPs and the Kenya potato and cassava CARPs, farmers inspired other farmers and farmer groups from local and surrounding areas who were not involved in the platforms. Seeing how the platforms impacted fellow farmers, others came to learn lessons from them. In this way, when farmer groups organised themselves, it strengthened the resilience of their communities and the sustainability of the entire platform, providing new impetus for engagements to continue beyond the official life of the programmes.

Through the pig CARP, a pork processing incubation centre was also established, which is enabling farmers and students to collaborate and continue to develop their skills and value-added products for which they are in the process of gaining certification. In the pig CARP, several early adopter farmers became role models to others and have grown their incomes several times over by employing innovative strategies. Odoch Geoffrey is one of several farmers who moved from selling live pigs to slaughtering and processing the pigs himself to selling them at a greater profit both as fresh pork and as deep-fried pork. From his pork sales, he invested in planting maize and sunflowers to feed his pigs and engaged additional outgrowers to also produce sunflower seeds. An abundance of pig manure from the IMO deep litter floor provided rich compost for growing these crops. Then with an abundant supply of seeds, Odoch struck an agreement to sell to local oil extractors but retain the leftover seed cake as pig feed. Now Odoch is working with local youth to extend this innovative scheme further, creating new opportunities for these youth in the process.

Also in the Nairobi cassava CARP, certain farmers became role models, training and inspiring others in and beyond their communities. One farmer, 'Mama' Jane Ndela, attended trainings at the University of Nairobi and progressed from a subsistence cassava farmer to becoming a baker, formulating cassava flour and developing new confections. Through interaction with a WhatsApp group connected to the CARP, she was supported to become a baking expert and was able to support her family through her enterprise. She gained recognition from the Taita-Taveta County which engaged her in training other women's groups in enterprise development and value addition. From her sales, she invested in a grinding mill to produce cassava flour. As part of the CARP, she was trained by the Seed Enterprise Management Institute of the University of Nairobi where she learned to intercrop cassava with cowpeas and fortify her cassava flour with dried cowpea leaves for improved nutrition (RUFORUM, 2020).

Building farmers' capacities in the utilisation of cassava leaves as a vegetable was important in Kenya where micronutrient malnutrition is high. Prior to the CARP, communities were not using cassava leaves because in addition to their high protein, vitamin, mineral and fibre content they also contain cyanide and other toxins. In times of drought, however, cassava leaves are often the only vegetables available. The CARP trained farmers in techniques such as selective picking of leaves to reduce their toxic content for utilisation in hard times. The project aimed to develop products from cassava leaves with uniform chemical and microbial qualities. Combining different methods of fermentation, dehydration and packaging, techniques were developed for using the leaves as dietary and vegetable supplements (RUFORUM, 2020).

Thanks to such capacity-building efforts with farmers, when the COVID-19 pandemic hit, as lockdown regulations limited mobility and disrupted food markets, farmers found themselves better equipped to handle the disruptions because they had better knowledge of how to utilise their local resources. CARP Principal Investigator Professor Agnes Mwang'ombe wrote:

> As many of the world economies slide into recession, conversations of farmer preparedness to adapt their food systems and rural economies to new realities has increasingly become apparent with a focus on supporting smallholder farmers' response, recovery and resilience. Resilient crops such as cassava have provided smallholder farmer households a buffer for food security in the heat of the current crisis in the coastal region as well as in eastern Uganda.
>
> *(RUFORUM, 2020)*

In the pandemic, peoples' cassava gardens provided a consistent food supply and bolstered food security through all the disruptions. Taking these lessons to heart, the Taita-Taveta County Government adopted cassava as a priority crop and partnered with the University of Nairobi to allocate 40 acres of land for clean cassava seed production, boosting capacity to produce much sought-after clean planting materials through tissue culture and minisett technologies. Mama Jane Ndela was at the forefront of community efforts to mobilise cassava processing in response to the crisis, developing new cassava recipes including cassava-baked fish balls, value-added crisps and cassava beans, and training her fellow community members. Through such innovative and entrepreneurial adaptations, the community gained a new understanding of the value of cassava (RUFORUM, 2020).

Centres of leadership to centres of excellence

The experience of one of the earliest CARP projects, the Malawi fish CARP, demonstrates how from the platforms' limited scope of engagement a much wider impact can grow. As the multi-stakeholder platform drew focused attention towards collaborating on key value chain issues and constraints, the university was able to align itself to the needs of the system. The CARP platform activities consolidated knowledge and expertise, which could then be harnessed and put forward in new institutional linkages with the New Partnership for Africa's Development (NEPAD) of the African Union, and in a new World Bank Centre of Excellence in Aquaculture and Fisheries. The CARP had positioned and propelled LUANAR into a Center of Leadership in fisheries science, from where they proved their leadership and showcased that investing in innovative research can deliver development outcomes.

BOX 5.3 FROM CARP TO CENTRE OF EXCELLENCE

Professor Emmanuel Kaunda (2021), the former project Project Investigator, described the ripple effect of the project:

> We had proposed through the CARP to transform fish production in Malawi and set a target of producing around 700,000 kilograms of fish per annum. By the time the CARP finished, we had surpassed our target, with farmers producing 1,500,000 kilograms of fish. We were able to link up not just with communities but with farmer organisations. This was key to the sustainability of the project. Even though the CARP is finished, that organisation remains very strong. It is still organising farmers and mobilising resources, which means the farmers are not left without help.
>
> At the same time, it was key for me that we were able to tease out the challenges the farmers were facing. We noted that they faced the challenge

of not having proper feed. Everyone was saying that in Malawi our fish doesn't grow. We put together a proposal and got a grant from DFID to benchmark the feed. We went to the private sector and did research on their farms and were able to show that with feed from Zambia or China, our fish was growing three times as much as with our own feed. The issue was not the fish. It was the feed.

Then we mobilised through the CARP programme to influence policy, and also to influence the private sector. It is because of our research that Malawi started importing feed. Our next phase is to ensure that the feed is manufactured in the country. Because of the CARP programme, we were able to reach the policy sphere and our results were able to inform the African Union. We became the centrepiece of organising and coordinating what we call the Aquaculture Working Group. We influenced policy for the whole of Africa to the extent that the Africa Strategy on Fisheries and Aquaculture was mostly born through our work and partnerships with other organisations.

Because of those activities, we were ready to apply to be a World Bank Centre of Excellence in Aquaculture and Fisheries. We became one of 24 Higher Education centres of excellence which built international recognition for LUANAR and for our CARP. What it means for us now is that we get students from Sudan, Ghana, Nigeria, Uganda, Zambia, Tanzania – from all over Africa. We are able to reach out and build capacity for the region. This is all because of what we did with the results that we got from the CARP.

The newer CARPs profiled in this book continue to build on the early lessons and achievements of the initial ones. As the platforms strengthen through continued iterations of collaboration upon collaboration, one thing leading to the next, the impacts continue to grow and multiply.

In the Namibia bush value chain CARP, the complexities of managing ecosystems with communities were highlighted. Bush encroachment is a phenomenon thought to be due to changes in fire regimes, livestock grazing pressures and accelerated global warming as increased atmospheric CO_2 levels enhance the growth rate of plants. This phenomenon is linked to climate change and land degradation, as ecosystems important to livestock grazing undergo changes and become thick with fast-growing bushes in response to changing weather patterns. Managing bush encroachment is a question of climate change adaptation. The management of the land itself affects the ecosystem and may contribute to bush encroachment which degrades the carrying capacity of the land and results in a degraded ecosystem. Other countries with large dryland areas dependent on livestock production are grappling with similar challenges.

For this reason, Namibia CARP Principal Investigator Professor Simon Angombe presented the project at the United Nations climate change meeting COP 27 held in Sharm el-Sheikh, Egypt, in November 2022. At the meeting, he connected with interested stakeholders from Botswana and Ethiopia, and also with World Bank officials interested in developing proposals to access funds from the Green Climate Fund to assist developing countries with climate change adaptation and mitigation measures.

Building on the CARP work, Angombe is working to establish a Centre of Excellence in the bush value chain at the University of Namibia. While there is a strong climate-driven interest in how to manage bush encroachment, and a lot of potential climate and biodiversity finance available, the highly technical mechanisms of climate finance have made it difficult for countries and institutions to access the funding. With a new centre of excellence, Angombe said, the university has an opportunity to strengthen its role as a development facilitator through multi-stakeholder collaborations where the technical expertise is also developed and mobilised to access climate funds. Through NILALEG, [3] supported by GEF through UNDP, the University of Namibia was tasked with the implementation of a national system for monitoring progress towards spatial targets in Multilateral Agreements (MEAs). The unit focuses on assisting the Ministry of Environment, Forestry and Tourism to develop indicators for reporting on commitments under the three Rio conventions. This system will streamline the Namibian reporting system, which in turn has the potential to attract further funding to support the establishment of a Centre of Excellence, Angombe said (Angombe, 2022).

Another opportunity has arisen to create a new East African Centre of Excellence in tuber research from the efforts of the potato and cassava CARPs operating in Kenya and Uganda. Jointly, these CARPs have highlighted important lessons in mobilising the multi-stakeholder platforms involving the universities, research and extension systems to produce disease-free planting materials and support farmers in creating systems for their uptake and distribution. Egerton University cassava CARP Principal Investigator Professor Richard Mulwa (2022) commented:

> This is something people are not putting much emphasis on, but we are seeing a threat for maize in our region and we need to think of long term starch substitutes. A Centre of Excellence in cassava, potato, sweet potato and yams can extend our work to have a higher impact in the region and raise the profile of crops that have been neglected but can be used for food security.

In Kenya, the CARPs, focused on roots and tubers, felt they needed to provide evidence of their work in a unique way focusing on challenging the country's dependence on maize. Mulwa, working with Nairobi cassava CARP Principal Investigator Professor Agnes Mwang'ombe and Egerton potato CARP Principal Investigator Professor Anthony Kibe as well as the TAGDev programme Secretariat at Egerton

University organised a National Forum dialogue with a provocative title: 'De-maizing Kenya'. The dialogue attracted at least 428 participants from within and outside Kenya, including stakeholders from other universities, TVETs, industry, research institutions including the International Institute of Tropical Agriculture (IITA), a variety of County Governments and National Government agencies and NGOs such as the Consumers Federation of Kenya. The discourse was provocative – with one national news daily's headline reading: 'We can't stop maize "reggae" without a fight' – yet also an important reminder of the role that universities have to play in providing evidence and raising consciousness of issues that affect humanity and society. This underpins the long walk towards alternative sources of starch in the region and demonstrates that change is a gradual process.

Universities as fulcrums of development

The stories in this chapter highlight the multi-faceted roles of universities serving as fulcrums of development that have been stimulated through the CARPs. We have seen over the last decade of implementing the CARPs that they provide a foundation for products and services spinoffs developed from research, taken up through student and community initiatives. The CARPs have also demonstrated that when university education becomes experiential and action-oriented, industry partners are attracted to higher learning institutions. Through the CARPs, these varied roles have been seen in rich and diverse ways, both large and small. In Sudan, a modest home garden vegetable and livestock project with rural women initiated a new path for the university to undertake community engagement. As project Principal Investigator Professor Muna observes, the project exposed young girls in the community to wider perspectives and a better understanding of education, a small spinoff that had not been anticipated previously. She observes:

> The young girls need role models. They tend to get married as young as 13 or 14. They need role models. Now we see that a positive message about education comes from our project. We can change the mindsets of young women so that they see education as a chance to broaden their horizons, find work and earn money.
>
> *(Elhag, 2022)*

At the University of the Free State in South Africa, the processes of helping small-scale wool producers access global markets and further develop skills and value addition generated rich spinoffs involving the private sector, including banks and unrelated industries seeking to apply their job creation model in the wine industry to provide employment in the low season. Through the project, the university's own experimental farm was upgraded to a demonstration hub where innovations ranging from the practical shearing, washing and grading of wool to spinoff industries in spinning, weaving and felt jewellery production have been pioneered. 'In essence, it is all about linking the best of the research, teaching and learning skills of staff and students to specific learning and development needs of society', reflected the university's Rector and Vice-Chancellor Professor Francis William Peterson (RUFORUM, 2022a).

The multi-disciplinary, multi-sectoral, multi-agency nature of the CARP model helped to forge relationships not only between the university and the communities but with other actors that were part of the platform. It also triggered linkages between those agencies themselves. These spinoffs resulted in policy change and advocacy being strengthened and gave voice to the communities. The platforms provided entrepreneurs with the opportunity to identify gaps, resulting in better access to services. They strengthened and empowered research and extension agencies to connect better with the communities and resulted in improved reach for other commodity value chains. The linkages to the university resulted in improved connectivity for communities to other departments and faculties in the universities. For example, the Uganda potato CARP drew interest from the national government, which provided supplementary funds through the Makerere University Research and Innovation Fund (Mak-RIF) to expand on the work of the platform. The platform has also drawn interest from the Multi-Sectoral Food and Nutrition Security Project of the Ministry of Agriculture, Animal Industry and Fisheries.

Not only did the CARPs strengthen university relationships with external partners but they also helped to strengthen connectivity within the universities themselves. Several of the CARPs drew investment in new infrastructure at universities, which attracted researchers from other disciplines. In some cases, the CARPs, such as the wool CARP in South Africa, also indirectly helped to stimulate universities to develop infrastructure and policies to support innovation and industrialisation on campuses. Through exposure to the CARP approach and facilities, researchers from other faculties developed their own interests in product development. In the Zimbabwe CARP, for example, machinery for the platform including grinders, groundnut shelling machines and a sunflower oil extractor are housed within the university's Innovation Hub and are available for the wider community outside the project to use. In the community, fragrant tamarind seed pods are another underutilised resource that researchers gathered and brought back to the university to experiment with novel uses. While there are well-developed value chains for tamarind in West Africa, it is not widely used in Zimbabwe. Colleagues from the biology department became interested in developing products such as yoghurt to utilise tamarind's natural richness in Vitamin C and other micronutrients.

Zimbabwe CARP Principal Investigator Ronald Mandumbu (2022) remarked:

> When the machinery came in, the university started to change its mindset and got excited about developing products. The CARP has been the catalyst for a lot of innovation – university-wide. When the biology department came in we collected the tamarind to test as a flavourant. In the communities, if the trees have a value and a use as products, then they respect the environment and that plays into the climate and the water cycle. So there can be benefits to the environment as natural resources are well managed because they are valued in terms of developing a product.

> Masau and tamarind are both dry area trees. Both have a high Vitamin C content. So we were asking people in the chemistry and biology departments if they could make Vitamin C tablets? The CARP is making sure that every department that wants to participate in product development is participating. It's a spinoff of CARP.

Strengthened relationships with the TVETs through the CARPs have spun off into wider efforts to strengthen extension services in Kenya and Uganda. An institutional strengthening initiative arose to address the weaknesses of extension and advisory services systems across Africa. The initiative provides a new model to improve last-mile delivery of technologies, innovations and management practices to emerging farmer groups through the training and deployment of Community Based Agricultural Advisors (CBAA). The CBAAs are TVET graduates whose role is to help train, support and broaden the influence of emerging farmer groups. With support from the Mastercard Foundation, RUFORUM engaged universities and TVETs in Kenya and Uganda, including Baraka Agricultural College in Kenya, to pilot the training and deployment of the CBAAs. In this pilot phase, TVET graduates were placed within emerging farmer groups that required technical backstopping to organise and institutionalise.

In the pilot phase, Baraka College provided training for 40 CBAAs who were deployed to support the delivery of extension and advisory services to smallholder farmers in the rural areas of Nakuru and Baringo counties. The model has shown some early successes. In one instance, a farmer group in Uganda was able to bring on board ten other farmer groups to create the Amuria Adoacerit Farmer's Cooperative within two years. The group has formalised registration with a fully-fledged Cooperatives Certificate from the Uganda Cooperative Alliance, expanded their micro-finance credit facility, developed value addition of hibiscus and improved the agronomy of their crop enterprises. Further, they have become attractive to other development agencies including the Government that have provided a borehole, distributed seedlings and other services. Some of these CBAAs have now been recruited by other organisations such as World Vision to assist in providing extension services to farmers in their field sites.

These are some of the illustrations of universities in motion shifting the narrative as development facilitators.

Conclusion

Each of the CARPs was essentially a seed that was given modest funding over a four- to five-year duration to generate a more lasting intervention and practically demonstrate a new model of the university as a fulcrum of development in society. The programmes were multi-faceted and so were the spinoffs they generated. Universities as facilitators of multidisciplinary, multi-stakeholder platforms emphasising practical action and experiential learning is a new value proposition that has emerged from the CARPs as seen in the glimpses of the practical, tangible examples of how it has worked in various contexts. In the more mature CARPs, such as the Malawi fish CARP, the platforms have catalysed and influenced regional and pan-African policy. In the newer CARPs, the ripple effects and wider spinoffs are still emerging. Each project has to carefully consider how momentum could be sustained beyond the life of the project funding. The lessons of the CARPs have shown that strong farmer organisations and strong working relationships among platform stakeholders already go a long way in and of themselves towards sustaining the momentum of activities and opening up new and unexpected avenues for universities to build further on what the CARPs accomplished.

Notes

1 TAGDev student enterprises have created 2,200 jobs.
2 He shares his story in this video: www.youtube.com/watch?v=Lz56doQGJMI.
3 Namibia Integrated Landscape Approach for Enhancing Livelihoods and Environmental Governance to Eradicate Poverty.

Bibliography

Angombe, S. 2022. Personal Interview.

Elhag, M. 2021, 2022. Personal Interview.

Fox, Louise and Dhruv Gandhi. March 2021. 'Youth Employment in Sub-Saharan Africa: Progress and Prospects'. Brookings Research Paper. https://www.brookings.edu/wp-content/uploads/2021/03/21.03.24-IWOSS-Intro-paper_FINAL.pdf.

Gbollie, S. 2023. Personal Interview.

Habiyaremye, A., Habanabakize, T., and Nwosu, C. 2022. 'Bridging the Labour Market Skills Gap to Tackle Youth Unemployment in South Africa'. *The Economic and Labour Relations Review*, 33(4): 786–805.

Kaunda, E. 2021. Personal Interview.

Lindow, M. 2020. 'TAGDev Stories'. PDF. RUFORUM. https://www.ruforum.org/sites/default/files/RUFORUM%20TAGDev%20Stories.pdf.

Mulwa, R. 2022. Personal Interview.

Onyango, L. 2023. Personal Interview.

RUFORUM. 2022. 'The Movement of Changemakers in Africa'. Video. https://www.youtube.com/watch?v=KZqbTMUJX6A.

RUFORUM. 2022a. 'Thought Piece'. PDF. https://www.ruforum.org/Triennial2021/sites/default/files/Thought%20Piece%20Issue%2028_Prof%20Francis%20Petersen.pdf.

RUFORUM. 2020. 'Voices from the Field: Transforming Cassava Production through Delivery of Clean Planting Materials and Business Development in Coastal Kenya'. Vol. 5. RUFORUM.

RUFORUM. n.d. 'Strengthening Universities' Community Engagement for Transformed Agriculture'. Video. https://ruforum.org/impact/project/strengthening-universities-community-engagement-for-transformed-agriculture/.

Rwelamira, J. 2015. 'Strengthening Farmers Organizations and Civil Society Organizations'. Background Paper, African Development Bank. https://www.afdb.org/fileadmin/uploads/afdb/Documents/Events/DakAgri2015/Strengthening_Farmers_Organizations_and_Civil_Society_Organizations.pdf.

6

FACILITATING A STUDENT AND COMMUNITY-CENTRED, EXPERIENTIAL APPROACH TO RESEARCH AND INNOVATION

Nancy Mungai, Duncan Ongeng, Patience Mshenga and Megan Lindow

Introduction

In both Egerton and Gulu Universities the Transforming African Universities to meaningfully contribute to Africa's Growth and Development (TAGDev) program was established as a wider initiative to develop models of higher education which recast universities as fulcrums for development in their societies (see Box 6.1). Community action research projects (CARPs) in the two universities benefited from and contributed to substantial agri-enterprise and community engagements that were already underway. Thus, the CARPs have played an important role in wider processes of institutional transformation focused on student-centred learning aligned with the needs of communities and industry. Both universities were founded on principles of experiential learning and community engagement. Egerton University, Kenya's oldest institution of higher learning, was founded in 1939 as a Farm School and was upgraded to a full-fledged university in 1987. In northern Uganda, Gulu University was established in 2001/02 with a mission to nurture community development in a region emerging from nearly 20 years of civil conflict.

Among all the different institutions where CARPs operated, both universities were outliers because of their comprehensive and pre-existing institution-wide focus on experiential learning and community orientation. The strength of these initiatives in both universities was a major attractor for the TAGDev programme which focused on piloting new models of higher education in Africa. As shown in previous chapters, CARPs in some other universities often struggled against a conservative institutional status quo and cultures resistant to change. While the CARPs in Egerton and Gulu Universities did not operate free of such friction and constraint, their experiences also demonstrate how far the approach can be taken

DOI: 10.4324/9781003387497-8

towards seeding transformation when it is adopted as part of wider intentional processes of institutional change. Building on the continued efforts of both universities to engage meaningfully with their surrounding communities, the TAGDev program articulated the necessary changes to policy and culture to put clear and systemic emphasis on experiential learning and community engagement in both institutions which required strong intentionality and leadership. In this particular context, the CARPs' multi-stakeholder platforms for practical engagement and experiential learning have generated a wealth of relevant new knowledge, practice and evidence to support transformation in both universities.

Both universities have experienced a paradigm shift to promote student-centred community engagement. The mandate of universities is usually three-fold, imparting knowledge (teaching, managing students learning activities), creating knowledge (research, scholarship and innovation) and transferring knowledge and skills to the community (extension and outreach). Through TAGDev, the following approaches have been shown to facilitate community-centred service learning and engagement:

1. Creating an enabling policy environment that supports community-centred engagement,
2. Co-creation of priority areas for community engagement,
3. Coordinated implementation programs and feedback mechanisms from the community,
4. Flexible engagement programs that generate mutual benefits for the university and community members.

Lessons from Egerton University

Egerton University is located in Nakuru County, Kenya, well known for its high potential in agricultural productivity. Since 2017, Egerton University has been implementing Community Action Research Projects in Seed Potato and Cassava Value Chains. The university was a key institution that trained agricultural experts for the region and consistently emphasised experiential learning and practical training in the curricula since its inception. As the student population grew over the years, however, many of the practical aspects of training could no longer be effectively implemented due to financial and human resource constraints. Experiential learning is crucial in agricultural studies. Only through direct experience can students gain essential skills in food production, farm management, value addition and other agricultural practices that are essential to local economies. The majority of Africans still gain their livelihoods through small-scale farming, and students must have direct experience working with farmers if they are to gain an understanding of the complex and multi-faceted challenges they face, encompassing climate variability, soil fertility, disease management, access to finance and markets and much more.

In 2014, before the CARPs were initiated, [1] the university gained support from RUFORUM to revive its farm attachment programme to enable students to gain direct experience with farmers. Agricultural students were attached to local farmers for an eight-week programme in which they performed activities such as completing baseline surveys of the farm enterprises, documenting the farmers' challenges and opportunities and helping farmers respond with innovative practices and interventions appropriate to their context. In addition, the Faculty of Agriculture updated its curriculum to further incorporate practical training across four-year undergraduate degree programmes. Students are given plots on the university farm where they grow major fruit and vegetable crops and complete practical work along every step of the value chain, including post-harvest handling, marketing and value addition.

Through the programme, a framework for experiential learning and working with communities was developed, laying the groundwork for deeper engagements through the CARPs. The CARP projects drew on postgraduate students across diverse disciplines and departments who then participated in all aspects of the CARP. The students helped with conducting baseline surveys at the beginning of the projects which was helpful in identifying the research gaps in seed potato and cassava value chains. The students also participated in field activities and establishment of demonstration plots with farmers, in addition to contributing their own specific research within the projects' ambit. The students were actively engaged in information dissemination sessions of the CARP projects including farmer training, seminars, conferences, farmer field days, exhibitions and case studies development among others.

The implementation of both CARP projects raised key challenges and learnings for the researchers and for the institution. Although the concepts of community engagement were familiar, both researchers and administrators were by no means experts, and learning curves were often steep. As agronomists, soil scientists, agricultural economists or disease specialists (plant pathologists) accustomed to working within disciplinary siloes, researchers had to quickly develop and put into practice critical new skills for facilitating multi-stakeholder collaborations. While it is true that universities have a mandate for conducting research and extension activities, and are indeed expected to work with communities, NGOs, local governments, the private sector and other partners, the CARPs with their deep focus on resolving practical challenges in specific value chains gave a much stronger impetus and impact to these collaborations.

The seed potato CARP, for example, arose in response to the challenges local farmers were facing in gaining access to quality seed potatoes. Identifying key gaps and constraints together, the researchers and farmers were able to implement training to enable the farmers to produce clean and healthy seed not only for themselves but also for other farmers. The co-learning through this intervention had a ripple effect, catalysing transformation within the community and the university. The structured nature of collaboration through the CARP platforms has achieved clear results and impacts that far exceeded previous research undertaken by the university. The CARP engagements have strengthened the visibility and partnerships of the university in the process. Both CARPs for example developed strong working relationships with the Baraka Agricultural College, a local TVET, which had not existed before. Building on the relationships established through the CARPs, the university has been able to contribute towards curriculum development and capacity building of staff at the TVETs, while students from Egerton have mentored TVET students in agri-enterprise development and entrepreneurship. The TVET students also benefited from experiential learning activities of the CARPs.

Through the CARP platform, relationships were also strengthened between the university and the local Nakuru County Government, which emerged as a key partner in the seed potato CARP. As the platform progressed the university was made visible to local government as a key reference point in terms of technical skills, capacity building and research (Lindow, 2020). Through the collaboration, the university was able to demonstrate the value of its research and innovation capacities. Nakuru County Government could appreciate the role of the university in widening farmers' access to clean seed potato and promoting an important local industry.

Within the university, lessons from the CARPs helped to inform the development of new policies and practices to bring community engagement into the day-to-day life of the institution. Drawing on the experience and lessons of the CARPs, a new community engagement strategy was implemented to provide the university with a set of institutional-level guidelines to support and coordinate community engagement practices across all the different faculties and operations of the university. CARP team members were instrumental in helping to draft the new strategy, sharing their experiences of building engagements, synergies and impacts with communities and giving lessons on how to ensure these engagements are generative.

The CARP has introduced a different orientation to the research process and demonstrated that this can be successful. While previously more traditional research efforts involved developing innovations and taking them to communities, the CARP approach requires an attitude of collaboration and responsiveness to the needs of communities. Involving communities in every aspect of the research process gives them a stake in the project outcomes. The experience gained through the CARPs has shown the university that it can influence the production of seed potato and the whole potato value chain. It has opened new avenues for sector-wide collaboration in marketing and value addition. It has helped farmers to see new business opportunities in their activities, and to see the university as a partner that can help them move forward.

The deliberate effort required to establish the CARPs and promote their thriving has also helped to revolutionise learning within the university. When students work with professors and with communities, they gain practical skills, and their classroom learning is enhanced. Working with farmers and learning from them also helps students to develop essential skills in communication, creativity and innovation for the 21st century that classroom learning cannot provide. When students are out in the field collaborating with different stakeholders, they gain confidence, learn to think critically, develop a better understanding of the challenges facing communities and are better able to innovate and adapt solutions that are appropriate to the context.

Learnings from the CARPs have also prompted a university-wide curricula review to put greater emphasis on practical, student-centred and problem-based learning. This deliberate, step-by-step effort has come with challenges and opportunities. For example, it has required substantial capacity building and reorientation of staff towards working with communities. It has required a mindset shift among staff and students alike. Students can no longer expect to just sit in the lecture hall. The aim of bringing community engagement into the day-to-day life of the university has provided new impetus for lecturers to incorporate community engagement in ways that support student learning. If each lecturer in every faculty – whether in medicine, commerce or nutrition and dietetics – integrates even a single aspect of community engagement into each course they deliver, students will gain experience with community engagement that spans their entire academic career.

As all of the university systems and processes become increasingly aligned with the goals of experiential learning and community engagement, the impacts and benefits of this approach grow clearer. Things are no longer business as usual. Students must now demonstrate that they are able to bring about change in communities and in workplaces. As the engaged approach of the university becomes more visible through the impacts of students and lecturers working with communities, the value of the university likewise becomes more visible to government, industry, communities and other partners.

This is expanding the scope of the universities' collaborations with industry and other partners. In the past, it was very difficult to interest the private sector in working with universities. However, as the university continues to develop a new model of community engagement, the private sector is becoming more interested, and collaborations are growing. Industries are now reaching out to the university to collaborate.

Such collaborations with the private sector will open up new avenues for students to gain skills and employment, and for the university to produce more industry-ready graduates with sought-after skills and experience. Such collaborations will also demonstrate the importance of university-led research and innovations to the local economy and to the nation. With communities and industries taking up the innovations of universities, the universities demonstrate their potential to contribute to economic growth.

All the different facets of collaboration and the spinoffs these have generated have helped many young people gain exposure to a wider set of possibilities for agriculture than they might have otherwise thought possible had they kept to a more traditional or conventional education path. Many primary and secondary school groups visit the campus to learn about agriculture and are able to see for themselves that agriculture is not just about digging with a hoe on the farm. They are able to see and appreciate the diverse applications of agro-processing, value addition, marketing, digitisation and information technology to agriculture and the food system – although many barriers to entry remain, including access to land, finance and markets.

Nevertheless, success stories are emerging. Graduates of the CARPs are now finding success as entrepreneurs and job creators. The university has learned that it must be deliberate in its embrace of experiential learning, community engagement and entrepreneurship. Through the exposure, practical experience and entrepreneurial training that such an approach provides, students are better supported to follow their natural drives to grow, learn, create and develop. They are gaining the skills and confidence to help farmers commercialise and even create their own businesses. In the past, the format of education at the university stifled these drives. Now, by the time many students graduate, they are already clear about their goals and dreams and are well on their way towards achieving what they want to do in life.

Gulu University: marrying scientific and indigenous knowledge to enhance the value proposition of universities

Gulu University in northern Uganda was selected as the other pilot university for the TAGdev programme, thanks also to the institution's legacy of community engagement. The university's community engagement model developed in response to its particular context as a post-conflict university. When the university opened its doors, many of the founding leaders and academics were fresh and inexperienced, coming from the surrounding communities, having suffered directly from the conflict. These individuals were passionate about creating a development-oriented institution to contribute towards a peaceful and prosperous new beginning in the region.

The university's founders were highly motivated to prove that their new institution could make a difference to communities in such challenging circumstances. Poverty and food insecurity were rife, and many local community members had endured unspeakable traumas from the violence as well as from living in camps for the internally displaced. The university founders' task was incredibly difficult: they had to face the reality of building up agricultural training from zero. The poor infrastructure and scarcity of resources forced them to think outside the box. In the classical models of agricultural training, imported from the United States and Europe, practical training took place on a university farm. However, the university had no means to establish a training farm and had to look for other ways to provide hands-on training. As an experiment, the university arranged for students to train directly in the surrounding communities with local farmers. In hindsight, this provided the opportunity for a fundamentally new approach to training that was in fact much better aligned to the local needs and context and set the stage for the two CARPs that were introduced later under TAGDev.

The key question at the time was: how can the university understand the failures and successes of the current approaches to training in relation to its own post-war context? Most colleagues were themselves products of ivory tower universities and did not necessarily question the traditional approaches. It was the only model they knew. Yet when those traditional approaches were not having the desired effect on research uptake, the academics realised they needed to change their mindset.

Academics asked themselves: what is the relevance to society of a university graduate trained in the traditional model; and what is the role of a university in a post-conflict environment in helping people get back on their feet? Investigating these questions, they realised that from a community perspective, the model of training on a university farm would not add value, because having students do their work on the farm keeps the university detached from the community. It was clear the university needed to engage directly with communities, both to contribute towards development and so as not to be seen as an elite world apart. There would immediately be a disconnect between the realities of local subsistence smallholder farmers and a university farm. The university had to work within the realities of the lands and resources available to local farmers if they wanted to help them advance.

University leaders also realised they would need to sensitise students to the conditions of local farmers. The costs of university are unaffordable to most rural families and, thus, the system naturally favours those from affluent and urban backgrounds who often have little orientation towards farming. Often even the lecturers were ill prepared for working with communities. They have hardly ever seen a cow and think that apples come from the supermarket. It was found that upon completion of their studies, many students favoured urban employment over employment in agriculture in the rural areas and that those who did enter the rural workforce were ill prepared. Having been trained in settings where cows were milked by machine, they had no understanding of how a milking cow can kick! They had little exposure to the indigenous knowledge held in communities and little understanding of its importance as a complement to their scientific knowledge.

To sensitise students and show its relevance, the university embraced a philosophy of community engagement from the beginning. The aim was to expose students to the intricacies of the rural environment and to help them learn to adapt and behave appropriately. The principle behind this approach was to train students in a way that married scientific and indigenous knowledge.

Similar thought processes informed the university's research agenda. Again, academics questioned the kinds of research that would be appropriate in an impoverished, post-conflict setting. If the university was to derive its research mandate from the local community, it obviously needed to interrogate the goals of its research agenda, which in traditional academic settings by default prioritised publications in prestigious academic journals. Should the aim of the research be to garner publications or to help the community in practical, tangible ways? At the time, these two goals seemed at odds. Again, following its philosophy of community engagement, the university had to be intentional about putting community-engaged research first. Academics and students alike needed new mental models to engage with communities and undertake research in a collaborative manner that supported community wellbeing.

University leaders chose to put students at the centre of the new community engagement model. This was also intentional. They reasoned that since students were being trained as future community development practitioners, they should be the ones to lead the exchanges between communities and the universities, bringing scientific knowledge to communities and indigenous knowledge to the institution, with their efforts backstopped by their lecturers.

Because the surrounding communities were deeply impoverished and relied almost entirely on subsistence agriculture for their livelihoods, agri-enterprise development became a natural focus of collaboration. In response to community needs, the university developed an agribusiness innovation model to guide research and community engagement to support the development of new enterprises to create jobs and improve farmers' incomes in communities. Thus emerged a student-centred approach of marrying science and innovation from the university with indigenous knowledge from the communities in order to develop rural livelihoods and equip students as agri-entrepreneurs and job creators.

Although agriculture is often not considered by young people as a 'sexy' career choice, it is nevertheless where the opportunities are. As experience with the agri-business innovation model started to demonstrate, students found that they could succeed as agri-entrepreneurs. Students were mentored through a process of identifying opportunities, creating business plans and running their businesses supported by seed funding. With support and guidance, a number of students were able to successfully run their own businesses, accumulating capital, employing people and earning an income on a par with a civil servant. As the success stories became more visible, greater numbers of students could see a viable path for themselves as entrepreneurs and find the courage and inspiration to pursue it.

These ongoing processes of student-centred learning, agri-enterprise development and community engagement provided a strong foundation for the two CARPs to build upon. In turn, the CARPs then contributed towards further strengthening engagement models to be taken up university wide. The practical focus of the CARPs in the rice and pig value chains greatly aided these processes. Prior to the CARPs, most community action research efforts had not focused adequately on bringing about economic development and wellbeing in communities. Under the CARP approach, farmer livelihoods became the central focus of multi-pronged practical activities led by a co-determined and continuously evolving research agenda.

Both CARPs have catalysed economic empowerment in the communities. The rice CARP was built upon a pre-existing government initiative to promote upland rice production in the region. Without a multi-stakeholder platform geared towards livelihoods, however, the initiative was underperforming. Only when the CARP was initiated and brought together many different actors for deeper knowledge-sharing and engagement did the situation actually begin to change in the communities. Farmers who had previously been growing rice at a subsistence level underwent metamorphosis, forming cooperatives and multiplying their incomes.

Farmers in collaboration with students and researchers developed powerful new income streams, including parboiled rice and clean-burning charcoal briquettes produced from rice straw waste.

In the rice and pig CARPs, businesses have been developed and incomes have multiplied from the principle of truncating the value chain. Traditionally, subsistence farmers have solely produced raw produce. Lacking market outlets and knowledge of post-harvest handling and storage practices, their only option was to sell to middlemen for very low prices. The multi-faceted CARP learnings and engagements have changed this paradigm, making it possible for farmers to organise themselves, pool their resources, drive their own learning of new skills and practices and develop their own agri-enterprises, venturing into much more lucrative processing and value-addition streams. Technologies have been developed through the marriage of science and innovation and indigenous knowledge which have enabled farmers to truncate the value chain and claim more value from what they produce. Farmers who had previously been at subsistence level are now earning up to US$1,000 per month.

Through the CARPs, the university has learnt that if it is to make a meaningful contribution to community development, it must have a strong and practical presence in the community. The university must derive its research mandate from dialoguing with the community. The research must be participatory rather than extractive. Research must be undertaken in a manner that generates practical and continuous learnings for students, communities and institutions alike. The results of research must be generated, packaged and put into practice through the efforts of students.

This is a challenge when institutions and research practices are modelled towards publishing papers. Changing the academic culture and incentive structures requires rewiring the brain to think differently. New practices will often encounter resistance, and universities can be notoriously entrenched in the status quo. Academics are accustomed to comparing themselves to their peers. If their peers measure their progress in terms of the papers they publish, they may struggle to stay motivated in the hard slog of community work. However, the CARP approach has shown that academics can successfully engage with communities and publish their work. It does not have to be either or.

Above all, such transformations require strong leadership which is itself transformative. Leaders must be committed to introducing the necessary mindset changes at the levels of students, staff and communities. At Gulu University, as institutional leaders have directly observed the student-centred community engagement and agri-enterprise models bearing fruit, they have become champions of institution-wide transformation. Gulu University has experienced a positive disruption of higher education for the meaningful empowerment of rural people.

In the case of students, for example, the university faced an uphill struggle. Having worked so hard to secure a university education and seeing it as a pathway to upward mobility through traditional channels such as studying medicine or law, students were at first reluctant to get their hands dirty working in the fields with farmers. Students were wanting to have their nails painted, not to dig in the soil or build a pigsty.

At last, the proof came when the university tracked the performance of its graduates and realised that the students who were finding employment were the ones with dirty hands. Those students who kept their nails clean were not sought after by employers. They did not have the practical experience and skills that employers were looking for. Thanks to the evidence produced through the CARPs and experiences such as these, the university has been implementing an institution-wide policy of community engagement. New curricula and assessment practices introduced across the institution now reflect the culture of community engagement. Convinced of the value of community engagement, the leadership has been driving these policy changes.

This would not have happened were it not for initial experiments such as the creation of the farm attachment programme, which sowed the seeds for community engagement in the Faculty of Agriculture and showed a marked improvement in the quality and relevance of graduates trained in a new model. When the results of strong community engagement practices became evident, university leaders championed the uptake of the approach across the institution.

Conclusion

The CARPs from both Gulu and Egerton Universities benefited from the pre-existing community engagement and agribusiness models pioneered by the two faculties of agriculture. In turn, the lessons from the CARPs contributed further to the processes of strengthening the models of engagement being taken up university wide. In both institutions, the visible successes of the CARPs are now urging and inspiring other faculties of the university to pursue similar kinds of engagements.

One of the foremost challenges facing Africa as a continent is redundancy among the large youth population due in part to a lack of training and relevant skills. The experience of the CARPs in conjunction with the wider TAGDev programme has shown that it is deeply challenging, yet still very possible, to support the development of young people as entrepreneurs. Out of 81 TAGDev graduates from Gulu University so far, 15 young people are currently succeeding as entrepreneurs.[2] Together they are employing an additional 60 people in their enterprises. The top-performing entrepreneurs among them are earning an income of as much as US$3,500 per month. These early results demonstrate the impact of measures within the university to drive agri-entrepreneurship, for example through the provision of a student enterprise scheme that provides modest start-up funding and mentorship. With the support of leadership and conducive policies in place, these are measures any university could take without breaking the budget.

In the experiences of both universities, there are several preconditions which must be met in order to reap the fruits of the approach. The university must clearly define its purpose and orient itself to champion community engagement and development. Institutions do not need to compete with Harvard for publications and rankings. The leadership must provide the environment for the approach to succeed. This means ensuring that each student finds substantial opportunities for experiential learning and to get their hands dirty working with communities. In order to drive development, students must be able to connect their experiences on the ground with the skills and confidence to harness opportunities in rapidly evolving agrifood systems.

Notes

1 While the very first CARPs were initiated in 2010 as described in Chapter 1, the CARPs were only piloted in Egerton and Gulu Universities in 2017 as part of the wider TAGDev programme.
2 The others are mostly either in formal employment or in continuing studies.

Bibliography

Lindow, M. 2020. 'TAGDev Stories'. RUFORUM. https://www.ruforum.org/sites/default/files/RUFORUM%20TAGDev%20Stories.pdf.
Mshenga, P. 2022. Personal Interview.
Mungai, N. 2022. Personal Interview.
Ongeng, D. 2022. Personal Interview.
Sharma, Rachita. 2022. 'Africa Is the Next Big Skills Market – But Youth Need Education and Opportunities'. Africa Portal. https://www.africaportal.org/features/africa-is-the-next-big-skills-market-but-youth-need-education-and-opportunities/.
RUFORUM, 2022. '2022 TAGDev Video with Subtitles'. Video. https://www.youtube.com/watch?v=W5j8CB9_4Fk.

PART II

The specific impacts of each of the fifteen Community Action Research Platform projects implemented 2016–2022

Part I has broadly described and used stories from the CARP (Community Action Research Platform) projects to show their potential to strengthen the relevance of university graduates, of university research, and of university systems to support smallholder agriculture and inclusive and sustainable national growth. Part II provides more specific detail of the implementation and impact of each of these CARPs. This section will be useful to students, rural development practitioners, educators, and agri-food system scientists in a range of disciplines. The chapters are written by the Principal Investigators with their implementing teams. Each chapter highlights how they approached establishing the platforms, working with smallholders to identify their problems, using students to address specific research, and develop innovations to address those problems across the full value chain.

Each CARP includes university faculty to lead and supervise post-graduate students to undertake the research assisted by undergraduate, technical, and vocational college students. All the field research was working with farmers on their farms and all laboratory work and product development had the results tested with the farmers and small entrepreneurs working along the value chains. It was truly participatory action research. But the approach went much further. Because it was addressing farmer problems it needed to look at the commodities within the entire value chain. There is no point in increasing yield for products you cannot sell or do not eat. This meant that the research teams had to be multi-disciplinary. They would not address every aspect along the value chain but would need to identify all the bottlenecks and put together teams of scientists to address several different aspects from soil fertility to processing; from animal husbandry to markets; from

DOI: 10.4324/9781003387497-9

nutrition to biodiversity and climate adaptation. Equally important, the community needed more than only the universities to help them. They needed all the key stakeholders in those value chains. Thus, the CARPs needed to be multi-agency. In addition, they also focused attention on developing products and small businesses for the students and the farmers to engage in. The CARPs became integrated into their societies and have helped strengthen farmer linkages to services and policymakers. The exposure provided experiential learning for faculty, students, farmers, cooperatives, and the companies, extension agents, local governments, and other agencies. The platforms built relationships and have shown their potential for sustainability.

Every CARP faced similar and yet different issues. All grappled with the demands of initiating and facilitating the innovation platforms. The transaction costs are high and, in the beginning, this did not seem worthwhile to the academics leading the projects. In the end all have found it to be very rewarding and all see the real value of using the approach for universities to have meaningful impacts in their local societies. Most of the CARPs were four-year projects (slightly extended by COVID pandemic complications) and included both doctoral and master's students with a significant role for the multi-agency platforms. The CARPs in the Sudan were shorter and less complex, training only master's students, but have addressed key issues. All of the CARPs had remarkable success in using research to improve the lives of smallholder farmers and rural communities, in preparing competent and creative graduates, and in raising the visibility of their universities.

The chapters that follow are loosely grouped into those value chains which are primarily focused on commercialisation, those that focused on the role of roots and tubers in food security, animal value chains, and then the CARPS which specifically include supporting farm systems and commodities that help to address climate change.

7

ENHANCING AGRIBUSINESS RICE CLUSTERS AND MARKET LINKAGES FOR FOOD SECURITY AND INCOMES IN NORTHERN UGANDA

Basil Mugonola, Apollo Kasharu, Jimmy Lamo and Daniel Ogwang

Introduction

Rice in Uganda has become an important dual-purpose crop that serves both as a food security and income crop to the majority of smallholder farmers (Ssebagala et al., 2017). The country's demand for rice outstrips the quantity produced, which makes Uganda a net importer of rice. Given the relevance of rice, the government has over the years promoted the production and dissemination of NERICA (New Rice for Africa) rice varieties. The enhanced popularity of the crop stems from the purposed breeding programs of the Uganda national research agencies and the government of Uganda's rice strategies: the National Agricultural Research Organization (NARO) that promoted improved high-yielding upland rice varieties and the Uganda National Rice strategy (URS) that sought to attain self-sufficiency in rice for the country (MAAIF, 2010). The National cereals program at the National Crops Resources Research Institute (NaCRRI) released high-yielding rice varieties with yields of about 4.5 tons per hectare (Lamo et al., 2017). Despite the widespread adoption of the improved upland rice varieties, yield differentials persisted between farmers' fields and research stations. Farmers obtained less than one third of the potential rice yields (Fungo et al., 2013).

This project "Enhancing agribusiness rice clusters and market linkages for food security and incomes in northern Uganda" worked with smallholder farmers (from Gulu, Nwoya and Amuru districts) who are generally geographically dispersed and produce limited quantities of rice that are usually below acceptable quality thresholds. Post-harvest losses are very high, significantly accounting for over 40% of the produce lost (Ssebaggala et al., 2017). The marketable surpluses were low and frequently bulk marketing was not possible because farmers' groups/producer organizations were either weak or non-existent. These scenarios escalated the transport

DOI: 10.4324/9781003387497-10

and other transaction costs for gathering the paddy rice and eventually eroded the farm-gate margins that accrued to the smallholder farmers.

To address these bottlenecks, a community action research programme on rice (CARP) was conceived to provide a platform for brokering functional and market-responsive partnerships to respond to the challenges. In the rice CARP, the agribusiness clustering approach (concentration of producers, agribusinesses, inputs stockists and institutions to address "wicked" problems and seek opportunities in the rice value chain) led to the improvement of market access conditions through collective marketing for inputs and outputs by smallholder rice farmers, improved productivity, competitiveness and efficiency. The improved performance in productivity and collective resolve of partners spurred the development of other products for both humans and livestock, initiated the utilization of rice husks for energy briquettes and enhanced market linkages in the rice value chain in northern Uganda. All the interventions embraced participatory approaches and action-oriented research methods including deliberate gender integration. The value chain actors including farmers, processors, mill owners and traders were part of every activity which further created a sense of ownership of the interventions and was key for adoption and sustainability in the rice value chain.

Moreover, the formation and strengthening of upland rice agribusiness clusters enabled the smallholder farmers to tap into the existing bulking stores and other infrastructure already present in northern Uganda (WFP.P4P, 2014). Our approach deployed Gulu University students and those of Technical Vocational Educational Training Institutions (TVETS) to work with various actors in the rice value chain. This provided opportunities, on the one hand, for unleashing both technical and functional capacities of the rice actors and, on the other, ensured that the divide in the higher educational value was bridged. Our approach was further premised on the concept of discrete support to encourage, nurture and form self-sustaining value chain Agribusiness Clusters (ABCs) in the rice value chain. Once these ABCs were deepened, actors were motivated to work towards competitiveness, vertical integration and upgrading activities and were also supported to embrace horizontal integration, building local networks and product development to serve different market segments and niches.

The rice CARP was implemented in four districts of northern Uganda (Gulu, Amuru, Nwoya and Omoro) and worked directly with producers' groups in these areas. In total, seven farmer groups – each group with an average of 150 beneficiaries – were impacted by the interventions. Further, other partners including Bobi Community Polytechnique, the International Institute of Rural Reconstruction (IIRR), the National Agricultural Research Organization (NARO), Chain Uganda, the Uganda Industrial Research Institute (UIRI), farmer cooperatives, rice millers and district local governments in the respective districts were involved in the rice CARP. A number of students from the TVETs, undergraduate, master's and PhDs were involved along with senior researchers. Seventy-five certificate students, four undergraduates (BSc Agriculture) students, nine MSc students and two PhD students[1] were involved in various research components of the rice CARP. The students' research areas involved participatory varietal selection in the rice variety trials, adoption of improved technologies in rice production, efficiency of production and market systems, production of energy briquettes from rice husks, production of parboiled rice and optimization of animal feeds from rice straws.[2]

Approach/methodology/setting

The Participatory Market Chain Analysis (PMCA) approach contends that improved market access is crucial if the competitiveness of the rural smallholder rice farmers is to be enhanced (Bernet et al., 2006). Promotion of collaboration along the commodity chain among various stakeholders has been found to be a promising strategy to (i) increase efficiency in the market chain by lowering the production and transaction costs which occur among the different value chain actors and (ii) enhance the value of the products and services created along the value chain, thus justifying the increased consumer sale prices.

The bottlenecks that surround the smallholder farmer (limited market access, reduced market opportunities, poor resource base and low geographic competitiveness) were especially targeted and turned into positives through carefully planned participatory market chain collaboration as depicted in the conceptual framework.

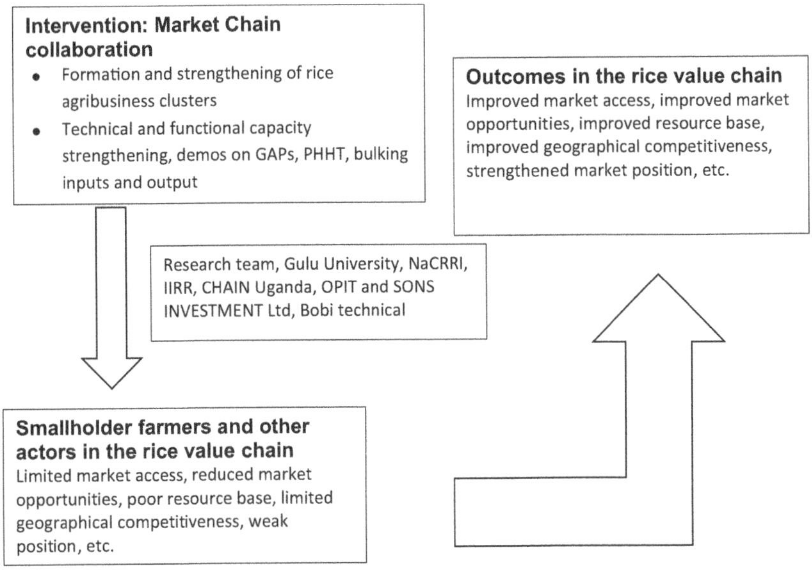

FIGURE 7.1 Conceptual framework and theory of change adapted from the participatory market chain analysis users' guide

To ensure that smallholder rice farmers leapfrog their surrounding constraints, a theory of change was instituted through action research involving TVETs, undergraduate and graduate students on one hand and a multidisciplinary team of scientists on the other, to bring out the desired change along an action research proposed impact pathway. By forming and strengthening rice agribusiness clusters concentrated around a miller who operates as a lead firm, the bottlenecks faced by smallholder farmers in accessing inputs, information, credit and output markets were addressed. OPIT and SONS INVESTMENTS Limited is one of the largest grain millers located in Gulu town with an installed capacity of more than 60 metric tons per day ability to mill different grades of rice and other cereals. This miller provides inputs and credit to their affiliated farmers in Gulu, Kitgum, Amuru, Nwoya and other districts in northern Uganda. One of the farmer cooperatives initially affiliated with the mill was the Pabbo rice grower's cooperative located in Amuru district. Other groups were strengthened following this model and linkages created with other millers including Payero rice millers in Gulu town. This action research project created the desired changes among the actors through the approaches of demonstration plots (learning plots), the capacity strengthening of actors along the value chain, and farmer education.

Results

The rice CARP "Enhancing agribusiness rice clusters and market linkages for food security and incomes in northern Uganda" sought to intervene in the rice value chain using a two-pronged approach. The first approach, focused on the production and supply side of the rice value chain, promoted Good Agronomic Practices (GAPs) and post-harvest handling technologies through on-farm demonstrations and learning plots to improve the production and productivity of rice. The second approach sought to create vibrant producer organizations and market linkages with various actors along the rice value chain for enhanced access to rice output/input markets. The synergistic combination of the two approaches has resulted in 'low hanging fruits' that point to the following three desired outcomes: improvements in rice production, improvements in productivity and market access for smallholder rice farmers. This was made possible due to the over 50% adoption of improved rice varieties (NAMCHE 3 and 5), adoption of correct planting densities due to the use of specially developed, locally fabricated rice planting equipment, use of drying tarpaulins to avoid contamination and enhanced access to marketing networks and the strengthened rice producer groups. The producer groups gained technical capacity (agronomy and post-harvest handling-maturity, moisture content) and organizational capacity that enabled them to increase acreage (by about 10%), productivity and to evolve into rice cooperatives and diversification into Local Seed Businesses (LSBs) with a number of producer groups

FIGURE 7.2 Experimental plots in farmer fields

selecting NAMCHE (1, 2 and 5) rice varieties for production of Quality Declared Seed (QDS) multiplication.

Following the technical capacitation in seed production, producer groups obtained certification from the Ministry of Agriculture, Animal Industry and Fisheries (MAAIF) to produce and trade QDS. To this effect, the producer groups have reported that they have higher profit margins in QDS than the rice grain. Other important results include parboiled rice, energy briquettes from rice husks (Nyuyki et al., 2022), rice noodles from broken rice, enhanced processes for technological and non-technological upgrading in the rice value chain (Akite et al., 2021; Akite et al., 2022).

FIGURE 7.3 Working in the fields

FIGURE 7.4 Rice growing field day

FIGURE 7.5 Farmers' association meeting

FIGURE 7.6 Farmers winnowing rice

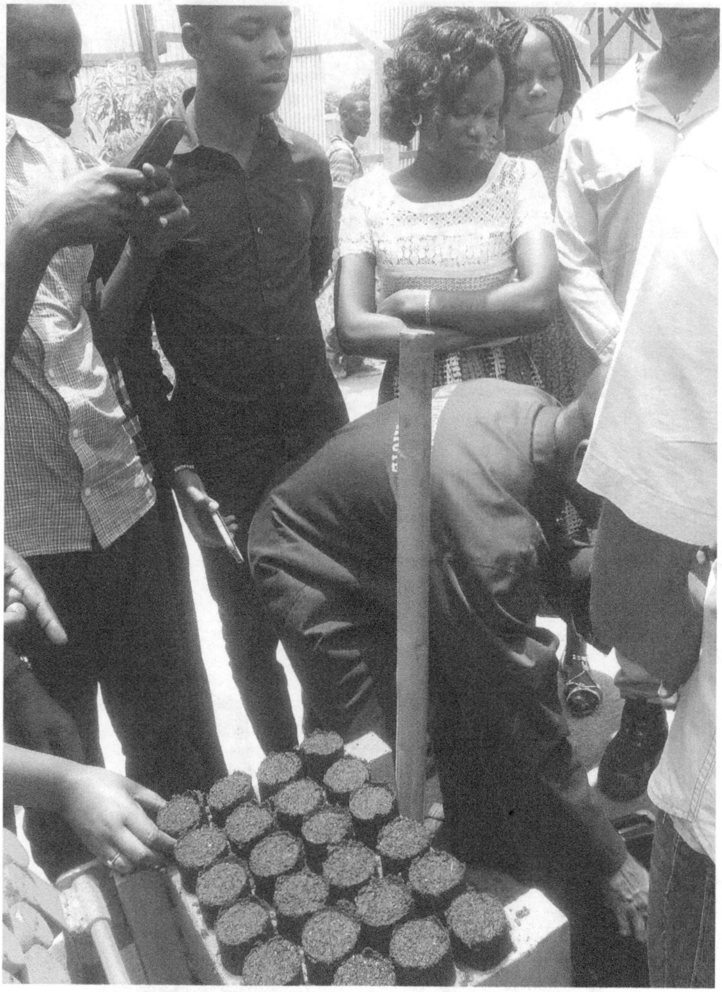

FIGURE 7.7 Students making charcoal

Postgraduate students who have graduated from the rice carp project with thesis titles

Impact

Through the rice CARP, we have demonstrated that this platform has the potential to contribute impactful actions to the wider community of value chain actors. The following are some of the highlights of the impacts contributed to through this rice CARP:

a) Uptake of innovations in communities by smallholder rice farmers. For example, increased uptake of improved rice seeds. The use of rice planters that were

TABLE 7.1 List of graduate students from the rice Carp project

Name of student	*Thesis title*
Edmond Nyuyki Mainimo	Determinants of willingness to pay for crop biomass derived briquettes in Gulu and Wakiso districts of Uganda
Lydia Nabantanzi	Drivers of adoption intensity of improved rice production technologies in northern Uganda
Muteti Francesca	Determinants of efficiency among rice producer groups and rice millers in northern Uganda
Prince Mpofu	Determinants of farmers' choice of improved upland rice varieties in northern Uganda
Joseph Aluong	Drivers of cross border rice trade between Uganda and South Sudan: the case of Elegu and Nimule border post
Ruth Ayado	Analysis of marketing forms for smallholder rice producers in northern Uganda
Christopher Atibo	Effect of malted iron-rich beans flour inclusion on sensory and nutritional properties of broken rice-iron rich beans composite flour for a complementary child food
Charles Bruce Nokrach	Effect of ensiling method on feeding value of rice straw-based silage for dairy cattle
Simon Peter Waigumba	Consumer's preference and willingness to pay for domestically produced rice in Uganda
Irine Akite	'Determinants of non-technological upgrading strategies for rice market improvements in Uganda'. *Scientific African*, 14(2). DOI:10.1016/j.sciaf.2021.e01030.
	'Estimation of profit efficiency of smallholder rice farmers in Uganda: A stochastic frontier approach'. *Journal of Agriculture and Food Research*, 8: 100315. DOI: 10.1016/j.jafr.2022.100315.

locally fabricated by TVET students that enabled farmers to plant rice in lines as opposed to the traditional practice of broadcasting. Certification by the Ministry of Agriculture, Animal Industry and Fisheries to produce Quality Declared Seeds (QDS) under smallholder local seed businesses. Diversified rice products including parboiled rice, fortified rice flour and rice noodles for home consumption and sale. The use of rice husks to produce high-density energy briquettes for cooking, thus reducing pressure on the use of woody biomass for charcoal. The use of rice straw to make animal feeds targeting urban and peri-urban farmers under zero grazing production systems. The ultimate impact of all these has been the stimulation of businesses within and without the rice value chain. Actors within the value chain have benefited from increased rice volumes and expanded trade networks within Uganda and South Sudan. While others outside the value chains benefited from trade opportunities in energy briquettes and rice straw for animal feeds. Community organizations have been strengthened and links for aggregation have improved marketing opportunities.

b) Created opportunities for more multi-disciplinary work and collaboration across the Faculty of Agriculture and Environment and with the Faculty of Business and Development Studies. It engaged within the Faculty of Agriculture with the Department of Rural Development and Agribusiness, Agronomy, Animal Production and Range Management, and Food Science and Post-harvest Handling.
c) Strengthened relations of Gulu University with the two technical vocational educational training institutions (TVETS) and charted a way to create mutual benefits for greater collaboration between post-secondary institutions. For example, the relationship between Gulu University, Bobi Community Polytechnique and Northern Uganda Youth Development Centre (NYDC) has been strengthened to the extent that there have since been jointly funded projects developed.
d) Established much stronger links between the university and communities which gives farmers and other actors access to Gulu University and vice versa. This further supports Gulu University in conducting action-oriented research, strengthening its community attachment programme and developing more relevant curricula that are responsive to societal issues and are demand driven.
e) Established co-operation within the National Agriculture Research System (NARS), other Universities, NARO, Ministries, Departments and Agencies (MDAs) and civil society organizations, non-governmental organizations (NGOs). Through this platform, we had opportunities to work with a number of government MDAs, especially on linking farmers to the best-improved varieties for the area. Links with NGOs and the private sector on relevant agri-commodity value chains have, for example, resulted in improved market opportunities. The parboiled rice initiative has been picked up by the Sasakawa Africa Association (SAA) and we hope to formalize a partnership to scale production and marketing of parboiled rice in Uganda.
f) Provided rural communities and university researchers with more sustainable links to a range of service providers and government policymakers – working together on a project that extends and deepens relationships for future collaboration in addition to creating an enabling environment for uptake and brings lasting benefits to communities. For example, two of the producer groups were able to attract government funding to buy tractors, build stores and install processing machinery.
g) Helped to transform the university for improved quality (supervision/links with outsiders/changed processes to speed up throughput of students etc.) and greater relevance (provided lecturers and administrators with the satisfaction of making a difference, seeing the university have an impact and increased publications etc.).
h) Raised the profile of the university in society – with local farming communities, with the research and/or development and government agencies and policymakers.

Conclusion

The CARP platform provides a window of opportunity that enables universities to broker Public–Private Partnerships (PPPs) that tackle multi-dimensional wicked (Strijker-van Asperen and Van Tulder, 2016) problems facing smallholder farmers. Consequently, universities are able to fill institutional voids that are of a public goods nature. The CARP platforms enable universities to broker and operationalize PPPs by attracting congruent partners, broadening the partnering spaces and a mutual contribution to shared value creation. Interventions that align with existing or emergent producer groups' business models are likely to succeed. For instance, through PPPs, two rice producer groups, one in Amuru and the other in Nwoya districts, were able to improve their organizational capacities and hence their external appeal for additional external support.

Further, the CARP platform provides a foundation for deepening the relevance of universities and other national agricultural research systems (NARS) in reaching out to communities, which provides graduate training and bridges the gap between University-TVETI educational value chains and connecting with the out-of-school counterparts. Future CARPs should broaden the partnering spaces further to enhance the spheres of influence and to bring on board government programming that captures the dynamic of agri-commodity value chain pathways.

Link to the film on the Rice CARP: https://ruforum.org/impact/project/gulu-rice-carp/.

Notes

1 See Table 7.1. The master's students have graduated, the PhD students are in advanced stages of their studies, having published a number of articles from their work. One of the PhD students is registered at Gulu University while the other student, who was the research assistant of the rice CARP, is registered at Egerton University in Kenya.

2 Research supervisors included: Professor Basil Mugonola (Department of Rural Development and Agribusiness, Faculty of Agriculture and Environment, Gulu University), Dr. Geoffrey Kawube (Department of Agronomy, Faculty of Agriculture and Environment, Gulu University), Dr. Christopher Mugagga (Department of Food Science and Post-harvest Handling, Faculty of Agriculture and Environment, Gulu University), Dr. Apollo Kasharu (CHAIN, Uganda) and Dr. Jimmy Lamwo (NARO, Uganda).

Bibliography

Akite, I., Okello, D. M., Lamo, J., Kasharu, A., and Mugonola, B. 2021. 'Determinants of non-technological upgrading strategies for rice market improvements in Uganda.' *Scientific African* 14 (2). DOI:10.1016/j.sciaf.2021.e01030.

Akite, I., Okello, D. M., Kasharu, A., and Mugonola. B. 2022. 'Estimation of profit efficiency of smallholder rice farmers in Uganda: A stochastic frontier approach'. *Journal of Agriculture and Food Research* 8: 100315. DOI: 10.1016/j.jafr.2022.100315.

Bernet, T., Thiele, G., and Zschocke, T. 2006. 'Participatory Market Chain Approach (PMCA)-User guide'. International Potato Center (CIP)-Papa Andina, Lima Peru. http://cipotato.org/wp-content/uploads/2014/09/003296.pdf.

Costa, S. J. 2014. 'Reducing food losses in sub-Saharan Africa: Improving post-harvest management and storage technologies for smallholder farmers'. An action research trial for Uganda and Burkina Faso. UN World Food Programme. https://documents.wfp.org/stellent/groups/public/documents/special_initiatives/WFP265205.pdf.

Fungo, B., Kabanyoro, R., Mugisa, I., and Kabiri, S. 2013. 'Narrowing yield-gap of rice through soil fertility management in the Lake Victoria Crescent agro-ecological zone, Uganda'. *African Journal of Agricultural Research*, 8(23): 2988–2999. https://academicjournals.org/journal/AJAR/article-full-text-pdf/64D04E836128.

KIT, Faida MaLi and IIRR. 2006. 'Chain empowerment: Supporting African farmers to develop markets'. Royal Tropical Institute, Amsterdam; Faida Market Link, Arusha; and International Institute of Rural Reconstruction, Nairobi. https://hdl.handle.net/10568/25011.

Lamo J., Tongoona, P., Sie, M., Semon, M., Onaga, G., and Okori, P. 2017. 'Upland rice breeding in Uganda: Initiatives and progress'. In *Advances in International Rice Research*, ed. Jinquan Li. InTech. DOI: 10.5772/66826 Published: under CC BY 3.0 license. Available from: http://www.intechopen.com/books/advances-in-international-rice-research/upland-rice-breeding-in-uganda-initiatives-and-progress.

MAAIF. 2010. 'Development Strategy and Investment Plan (DSIP)'. Ministry of Agriculture, Animal Industry & Fisheries. Kampala, Uganda. https://www.gafspfund.org/sites/default/files/inline-files/5.%20Uganda_strategy%20and%206.%20Uganda_investment%20plan.pdf.

Nyuyki, E. M., Okello, D. M., Mambo, W., and Mugonola, B. 2022. 'Drivers of household demand for cooking energy: A case of Central Uganda'. *Heliyon*, 8(2). DOI:10.1016/j.heliyon.2022.e09118.

Ssebaggala, G. L., Kibwika, P., Kyazze, F. B., and Karubanga, G. 2017. 'Farmers' perceptions of rice postharvest losses in Eastern Uganda'. *Journal of Agricultural Extension*, 21(2): 30–43. DOI: https://dx.doi.org/10.4314/jae.V21i2.3.

Strijker-van Asperen, Z. M. and van Tulder, R. 2016. *Wicked Problems Plaza – Principles and practices for effective multi-stakeholder dialogue*. The Partnerships Resource Centre (PrC). Retrieved from RePub. http://hdl.handle.net/1765/93223.

WFP.P4P. 2014. 'Connecting farmers to markets'. https://documents.wfp.org/stellent/groups/public/documents/communications/wfp287305.pdf.

8

ENHANCING SAFFLOWER PRODUCTION AND PRODUCT DEVELOPMENT FOR FOOD SECURITY AND IMPROVING INCOMES OF SMALL-SCALE FARMERS IN BOTSWANA

Vallantino Emongor, Bamphiti Tiroesele and Onkgolotse Moatshe-Mashiqa

Introduction

Safflower (*Carthamus tinctorius* L.) is a temperate crop that is cultivated in arid and semi-arid lands (ASALs) of the world (Emongor, 2010; Janmohammadi, 2015). Safflower is an underutilized, neglected, and minor crop compared to other oilseed crops despite its excellent adaptability to various climatic conditions, tolerance to drought, salinity and extreme temperatures, and the seeds containing essential fatty acids (oleic and linoleic) which are beneficial to human health (Emongor and Emongor, 2022; Mani et al., 2020; Moatshe et al., 2020a; Marang et al., 2022). Global warming is predicted to increase the frequency of droughts, floods, and heatwaves (IPCC, 2021). The rise in global temperatures may cause climate anomalies, resulting in crops encountering increasing abiotic and biotic stress combinations, which may negatively influence their growth, development, and yield (Mahalingam, 2015; Marang et al., 2022). The use of suitable crop species such as safflower (Figures 8.1 and 8.2) and a variety of cropping systems should be adopted in response to climate change (Acevedo et al., 2020; Hufnagel et al., 2020; Marang et al., 2022).

Drought stress causes reduced crop height, biomass, and yield. The quantity of available water for a crop is an important factor that determines yield. Drought negatively affects the already fragile food and agricultural situation in ASALs and impairs rural economies and socio-cultural structures. Due to erratic, unreliable, and poorly distributed rainfall, in combination with high temperatures (43°C) and evapotranspiration rate (1800–3000 mm per annum), water becomes the most restricting factor to agricultural production in Botswana and other ASALs (Emongor, 2009). Climate change is making rainfall less reliable and unevenly distributed which particularly affects these areas. Botswana is already vulnerable to drought,

DOI: 10.4324/9781003387497-11

FIGURE 8.1 Safflower at the elongation stage can be used for hay, silage or grazed

FIGURE 8.2 Safflower heads for florists, medicinal teas, or left as seed for oil

and this is likely to worsen with climate change. Therefore, it is important to research crops that are less dependent on regular rains and that tolerate salinity and extreme temperatures made worse by climate change. This means that it can be grown as both a summer and a winter crop (Emongor and Oagile, 2017). Furthermore, the research showed that safflower was relatively tolerant of pests.

Despite the potential of safflower and its drought, heat, cold, and salt tolerance, it has remained underutilized and a minor crop. This is largely due to a lack of information on its management, locally adapted varieties, product development, and reluctance of farmers to adopt a new crop. This CARP project was envisioned to explore the potential for safflower in Botswana because of its tolerance to these limiting abiotic factors and to engage with multiple stakeholders to overcome the lack of information. Growing a multipurpose, drought, saline, and temperature-tolerant crop such as safflower is expected to: (1) mitigate the effects of climate change in a semi-arid country such as Botswana; (2) improve food security, reduce reliance on food imports, and improve income levels of farmers in Botswana through the sale of safflower products such as oil, processed petals, cut-flowers, green leafy vegetable, and cake after oil extraction for livestock feed; (3) improve the livestock sub-sector through the availability of feed (seeds, cake, direct grazing, hay, and silage) and hence reduce feed importation, increase farmer income, and reduce poverty and hunger; and (4) improve health and well-being of Batswana of all ages through the use of safflower oil in cooking, salad dressings, and baby foods. Safflower has numerous health benefits due to the polyunsaturated and monounsaturated linoleic and oleic fatty acids. It is also used as a herbal tea and homoeopathic medicine. More information on production and potential use is available in the Safflower Production Manual (Emongor and Oagile, 2017).

Approach and methods

Once RUFORUM approved the safflower CARP project in June 2019, the Botswana University of Agriculture and Natural Resources (BUAN) began project implementation with a consortium of partners and institutions from the public and private sectors working with farmer groups, individual farmers, and community-based organizations. BUAN is the lead institution in the partnership working with the Department of Agricultural Research (DAR) and Extension Staff in the Ministry of Agricultural Development and Food Security, Ministry of Youth and Gender, University of Botswana, Botswana International University of Science and Technology, Kgatleng Brigades (TVET Institution), National Food Technology Research Centre, African Entrepreneurial Agency, the Healthy Families Foundation, Farmer Groups, and Individual Farmers. The project team members and collaborators have different expertise that fits well with the research objectives. Mentorships were both provided by the team members of the project and stakeholders who have been working with safflower and the localities where the project is implemented. The project was implemented in Kgatleng, Ramostwa, Kweneng, and Central Districts of Botswana. The Principal Investigator (PI), Co-PI, and MSc and PhD students in the CARP project have developed the following products: vegetables, herbal tea, seed, animal feeds (cake, meal, seed, roughage, hay, silage), and cooking oil which have been disseminated to stakeholders and are being produced with an increasing demand for safflower seed.

Project implementation was delayed by COVID-19 pandemic lockdowns. However, for the MSc and PhD students to complete coursework and defend their research proposals blended learning (online lectures using Moodle with face-to-face practicals) was introduced. Farmer training was done via WhatsApp. The research approach and conceptual framework for the implementation of the project are shown in Figure 8.3.

Results

The project has confirmed the optimum plant density that maximizes seed and oil yield, oil content, and fatty acid composition under ASAL conditions of Botswana to be 100,000 plants/ha at a spacing of 50 cm × 20 cm or 40 cm × 25 cm under dry land or irrigated farming (Emongor et al., 2013; Emongor and Oagile, 2017; Moatshe et al., 2020a). The fatty acid composition of safflower oil was linoleic (54–78%), oleic (11.1–23.4%), palmitic (7.0–16.1%), stearic (2.3–6.3%), arachidic (0–1.93%), iso-stearic (0–1.47%), and iso-oleic (0–1.6%) depending on genotype, plant density, and growing season (Moatshe, 2019; Moatshe et al., 2020a). Safflower oil is high in polyunsaturated and monounsaturated fatty acids

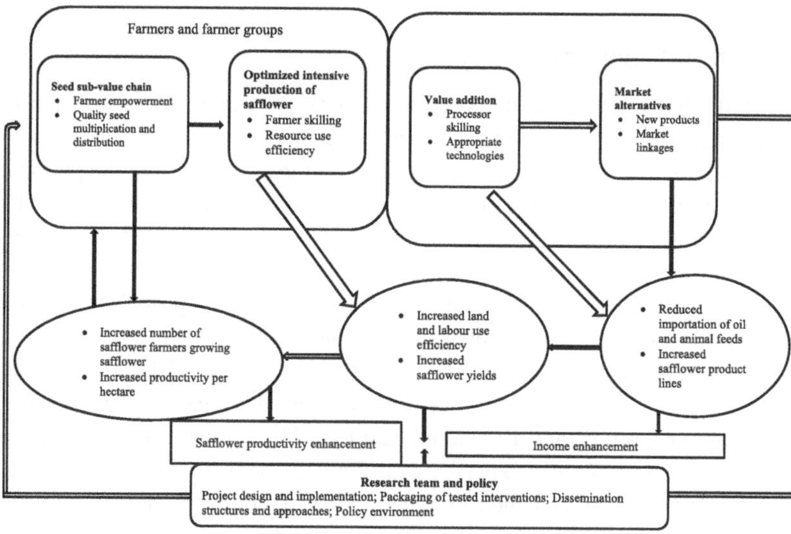

FIGURE 8.3 Implementing approach of safflower CARP

which are essential to human diet. Essential fatty acids have cardioprotective influence because of their antiatherogenic, antithrombic, anti-inflammatory, antiarrhythmic, and hypolipidemic roles. They also reduce the risk of cardiovascular diseases, cancer, osteoporosis, diabetes, and other health promotive activities due to their effect on lipoprotein contents, biological membrane fluidity, membraned enzyme and receptor function, eicosanoid production, blood pressure regulation, and mineral metabolism. The project also showed that safflower genotypes influence vegetative growth, phenological stages, yield components, seed and oil yield, and seed oil content (Oarabile et al., 2017; Emongor et al., 2017; Moatshe et al., 2020b; Emongor and Emongor, 2022). The optimal nitrogen (N) and phosphorus (P) fertilizer application rates to maximize safflower vegetative growth, seed yield and oil content under sandy loam soils of Botswana is 40–75 kg N/ha and 50 kg P/ha depending on genotype, growing site, and season (Mazhani, 2017 [see Table 8.1]; Kolanyane et al., 2022).

The results reflected that safflower cake after oil extraction contained crude protein (CP), neutral detergent fibre (NDF), acid detergent fibre (ADF), acid detergent lignin (ADL) and ash contents which significantly varied between 19.3–22.5, 54.6–61.2, 45.0–50.7, 18.0–20.8, 1.10–1.60%, respectively, depending on genotype and growing season (Phuduhudu et al., 2018; Kereilwe et al., 2020). The safflower seed mineral content significantly varied between 6.98–7.90 mg/g P, 10.68–12.91 mg/g K, 8.78–10.61 mg/g Ca, 4.45–4.99 mg/g Mg, 90–120 ppm Zn, 70–90 ppm Fe, 40–50 ppm Mn and 90–130 ppm Cu, respectively, depending on genotype and growing season (Phuduhudu et al., 2018; Kereilwe et al., 2020). Significant differences existed in the safflower seed oil content, DM, CP, NDF, ADF, ADL, and ash which varied between 26.13–42.17, 91.9–96.1, 16.3–19.1, 42.6–50.3, 39.7–48, 13.5–20.7, and 0.95–1.41%, respectively, depending on genotype and growing season (c, 2017 [see Table 8.1]). The safflower leaf DM, CP, ND, ADF, ADL, and ash significantly ($P < 0.05$) varied between 88.1–91.2, 21.1–27.7, 20.5–26.2, 26.5–32.7, 6.7–10.7, and 0.89–1.13%, respectively, depending on genotype and growing season (Phuduhudu, 2017 [see Table 8.1]). The leaf mineral content varied between 3.31–4.95 mg/g P, 56.13–66.54 mg/g K, 10.61–16.51 mg/g Ca, 3.91–4.92 mg/g Mg, 0.51–0.69 mg/g Na, 70–90 ppm Zn, 310–460 ppm Fe, 280–380 ppm Mn, and 6.3–8.3 ppm Cu, respectively, depending on genotype and growing season (Moatshe et al., 2020c). The above nutritional content of whole safflower seed, cake and leaves showed that safflower is an excellent animal feed (Kereilwe et al., 2020). The project also showed that safflower genotypes and planting time influence the agronomic traits of safflower, phenology, incidence, and severity of chilling injury (Moatshe et al., 2020b; Korononeo, 2022).

The results of this CARP project showed that 15 insect species belonging to eight orders were observed in safflower. Among the 15 species, ten were pests, four were predators and one was a pollinator (Keheng, 2022). Order Hemiptera had the highest number of species. *Thrips tabaci* and *Amrasca biguttula biguttula* were the most abundant insect species in summer and winter. *Helicoverpa armigera* and Aphididae species were identified as the most destructive pests of safflower (Keheng, 2022). Insect pest populations fluctuated along safflower phenological stages, but the most populated stage was flowering. The pests fed on all the upper parts (shoots) of safflower plants in the field, with leaves and capitula being the most affected parts of the crop. Even though insects were recorded in abundance, generally the impact of the pests did not significantly affect seed safflower yield in all five genotypes. This was attributed to the compensation ability of safflower plants. The highest diversity index in summer was recorded on genotype Sina (H'=1.47) and the lowest was recorded on PI-537636 (H'=1.32), while in winter the highest diversity was recorded on PI-537636 and Turkey (H'=0.94) and the lowest was H'=0.72 on Sina. The values suggested a non-significant difference in the diversity of insects between genotypes. The Sorensen similarity index also confirmed the similarity between genotypes as the Sorensen similarity coefficient varied from 96% to 100% in summer and varied from 80% to 100% in winter. The temperature indicated a non-significant quadratic correlation with the total population of insects in winter but a positive and non-significant linear correlation with rainfall and relative humidity. In summer, there was a positive and non-significant linear correlation between temperature and total population of insects, a non-significant quadratic correlation with rainfall and a positive curvilinear relationship with relative humidity (Keheng *op cit*).

Impact

Safflower is a crop being introduced into the cropping system of Botswana. The Principal Investigator (PI) Co-PI, MSc and PhD students, and stakeholders (individual farmers and farmer groups) in the CARP project have developed the following products: vegetables, herbal tea (from dried leaves, petals and roots), seed, animal feeds (cake, meal, seed, roughage, hay, silage), and cooking oil. The uptake of safflower by farmers has been excellent. A total of 787 farmers (204 males, 583 females) are growing safflower and selling the above safflower products with the exception of cooking oil. The project has empowered 168 women who suffered gender-based violence (GBV) during COVID-19 lockdowns in Rakops, Botswana. The women's group is selling safflower vegetables, seed, herbal tea, and animal feed to the community (Figures 8.4a, b, c, d). The Rakops women's group, Green Diamonds Co-operative Society, currently has a tender supplying weekly safflower vegetables to a primary school of 3,500 students. The crop can grow year round and is able to produce supplies in both the warm wet season and the cooler dry season without supplementary irrigation. This allows farmers to supply safflower products with more consistency which is important to stores.

The growing of safflower is mitigating the effects of climate change and food insecurity caused by low unpredictable rainfall and poor soils because it grows under rainfed conditions in sodic and infertile soils. It is producing good yields in environments where other crops such as maize, sorghum, millet, and sunflower

FIGURE 8.4A Herbal tea

cannot grow. Safflower production has improved food security, reduced reliance on food imports, and improved income levels of farmers in Botswana through the sale of safflower products such as petals, cut-flowers, vegetable, seed, roughage, and meal. Safflower has improved the livestock sub-sector through the availability of feed (seeds, cake, direct grazing, hay, and silage) and reduced the import bill of livestock feed and vegetables. It has the potential to replace imports of olive oil in high-end stores and to provide healthy cooking oil when expressed by farmers.

Due to these benefits, the Ministry of Agricultural Development and Food Security (MoA) has changed its policy on the Integrated Support Programme for Arable Agriculture Development (ISPAAD) to include safflower as one of the crops under

FIGURE 8.4B Leafy green vegetables in the supermarket

FIGURE 8.4C Seed for roughage and cake for animal feed

FIGURE 8.4D Safflower oil

the programme. The goal is to produce cooking oil, animal feed, herbal teas, pharmaceuticals, vegetables, cut flowers, and cosmetics for the country. In the amended policy, for every 4 ha supported by the government, 1 ha must be safflower. The government of Botswana is piloting this in Kweneng District in the 2023 growing season before rolling it across the entire country. The MoA has supported the training of 129 farmers in Kweneng District (12 November 2022) to grow safflower and recorded a training video (30 minutes) on safflower production, product development and processing which has been broadcasted on Botswana television thrice. Also, Botswana television interviewed the Principal Investigator and Co-PI on safflower prospects in the country and recorded a documentary which has been broadcast on three occasions (Botswana Television, 2022). The original 30-minute interview was broadcast live, which created awareness and interest across the country and ensured the sustainability of the initiatives established through the Safflower innovation platform.

Through this CARP project, capacity in agriculture has been built by training four PhD, six MSc, three BSc, ten TVET students, and 787 farmers. Further funding in safflower research was obtained by the Department of Agricultural Research

(Ministry of Agricultural Development and Food Security), from the Japan International Co-operation Agency (JICA), and Government of Botswana (GoB). The JICA project (US$300,000) is researching biodiesel development in various plants including safflower. The JICA project is sponsoring a PhD student (University of Botswana) with the Principal Investigator (PI) of the CARP project serving as a co-supervisor. The GoB project funded a PhD student, under the supervision of a PI, who investigated the influence of plant density and genotypes on phenology, agronomic traits, oil content, and composition of safflower. The student has graduated, and she is currently the Scientific Officer in the Ministry of Agricultural Development and Food Security. See Table 8.1 for the list of graduate students.

The CARP project has created opportunities for multi-disciplinary work and collaboration across departments and faculties within the university (BUAN), between universities (BUAN and University of Botswana), and between the university and other government parastatals (National Food Technology Research Centre [NAFTC]) and departments (Department of Agricultural Research [DAR]). The development of a safflower planter by the Kgatleng Brigades (TVET institution) has strengthened relations between the BUAN and the TVET institutions. The current CARP project through on-farm research and training of farmers, field days, and workshops has established much stronger links between the university and the farming community. The CARP trained the farmers and the opportunities resulting from safflower led to the formation of three co-operatives (Figure 8.5).

FIGURE 8.5 Farmer training in the rural areas

The CARP has facilitated farmers' access to the university and has helped them to access information on both safflower and other horticultural crops more easily. The emerging safflower research is demand driven. Through the CARP, farmers have organized themselves into groups and formed co-operatives (Letsema Co-operative, Kweneng North Horticultural Co-operative, and Green Diamonds Society). They are driving the university to engage further.

TABLE 8.1 Students involved in the safflower CARP and their research topics

Student Name	*Thesis and/or proposal for research*
Keheng, B.	Abundance and diversity of safflower (*Carthamus tinctorius* L.) Arthropods in Botswana. MSc Thesis, Crop and Soil Sciences Department, Botswana University of Agriculture and Natural Resources (BUAN). https://moodle.buan.ac.bw/handle/13049/554.
Kereilwe, D.	Temperature and duration of exposure on chilling injury of safflower. PhD Proposal, Crop and Soil Sciences Department, BUAN.
Korononeo, M. K.	Effects of planting date and genotypes on growth, development, yield, and oil content of safflower under irrigated conditions in semi-arid South-East Botswana. MSc Thesis, Crop and Soil Sciences Department, BUAN.
Maduo, O. K.	The influence of nitrogen and phosphorus nutrition on growth and yield components of safflower (*Carthamus tinctorius* L.). MSc Thesis, Crop and Soil Sciences Department, BUAN. https://moodle.buan.ac.bw/handle/13049/534.
Mazereku, C.	The influence of plant growth regulators on yield and yield components, oil content and yield, and biodiesel quality. PhD proposal (2020), Department of Chemical Engineering, University of Botswana.
Mazhani, L. C.	Effects of nitrogen and phosphorus on growth, development, yield and oil content of safflower (*Carthamus tinctorius* L.). MSc Thesis, Department of Crop and Soil Sciences, BUAN. http://moodle.buan.ac.bw/handle/13049/93.
Moatshe, O. G.	Genotype and plant density effects on growth, yield, and oil composition of safflower (*Carthamus tinctorius* L.). PhD Thesis, Department of Crop and Soil Sciences Department, BUAN.
Monyame, P.	Influence of nitrogen and phosphorus Fertilizer on safflower insects. MSc Proposal, Department of Crop and Soil Science, BUAN.
Mosupiemang, M.	Safflower growth, development, yield, oil content and composition as influenced by genotype and environment under on farm conditions. PhD Proposal, Department of Crop and Soil Sciences Department, BUAN.
Oarabile, P.	Effects of safflower (*Carthamus tinctorius* L.) genotypes on growth, development, yield and yield components, and oil content and yield. MSc Thesis, Department of Crop and Soil Sciences, BUAN https://researchhub.buan.ac.bw/handle/13049/91.
Phuduhudu, D.	Evaluation of the nutritional potential of safflower (*Carthamus tinctorius* L.) leaves, seed, and cake after oil extraction to be used as animal feed. MSc Thesis, Crop and Soil Sciences Department, BUAN https://moodle.buan.ac.bw/handle/13049/398.
Setshogela, B. P.	Evaluation of harvest time to harvest safflower petals for optimum petal and seed yield, carthamin, and carthamidin contents. MSc Proposal, Department of Crop and Soil Sciences Department, BUAN.

The safflower CARP gave opportunities for lecturers to be supervisors for MSc and PhD students which feeds into their key performance area of research and contributes to promotion and fulfilment of performance management system requirements of BUAN. Research-generated publications (MSc and PhD Theses, and papers in refereed journals) and outreach to the community through the safflower CARP have contributed to BUAN's vision of being a research-intense institution and improved the visibility of the university nationally, regionally, and internationally. Safflower CARP has raised the profile of the university with local farming communities through outreach and/or extension services, research, development, and linkages with government agencies and policymakers. The safflower CARP has also helped the PI to connect with the University of Botswana and Botswana International University of Science and Technology (BIUST) in joint supervision of MSc and PhD students and research in multidisciplinary disciplines such as chemistry and engineering in the development of safflower products such as biodiesel and cooking oil, which leads to sustainability and further development of new products through research.

Challenges

The major challenge was the COVID-19 pandemic, especially during the six-month lockdown period in 2020 when programme activities could not be implemented. The master's students could not complete their coursework in time which affected the commencement of their research. The other challenge was that farmers expected the project to provide all inputs during on-farm research including irrigation pumps and drippers, fencing the experimental fields, and all agricultural activities (ploughing, planting, weeding, and harvesting). The project supported some of these activities but not fencing and irrigation pumps and related accessories. There was a challenge of government bureaucracy in getting a TVET institution to work with an outside institution which delayed collaboration between the project at the BUAN with the TVET colleges. Reaching out to the government was made possible through support to some of the initiatives undertaken to assist women in difficult circumstances cut off by the pandemic. What was achieved with the women's groups helped to highlight the potential of safflower to both local and national government. This helped to overcome issues related to increasing the profile of safflower, which also triggered policy changes and attracted national media attention.

Conclusion

Safflower has the great potential to industrialize and diversify the economies of countries in the ASALs due to the many products that can be developed and commercialized ranging from healthy vegetable oil, pharmaceuticals, livestock feed, the food industry (leafy vegetables, colouring, herbal tea, and cooking oil), biofuel (biodiesel, ethanol, biogas), the textile industry (dyeing), the cosmetic industry, and the paint industry. These all need further research and investment to be able to fully understand the potential demand and address production and processing constraints. Implementation of effective and specific support policies, technological inputs, pricing, and marketing systems by governments in ASALs into safflower has significant potential to improve food security and health, reduce the import bill, and improve disposable incomes and livelihoods alleviating the poverty faced by many farmers. Interested readers can see the Botswana Television documentary on safflower produced in association with this CARP.

Bibliography

Acevedo, M., Pixley, K., Zinyengere, N., Meng, S., Tufan, H., Cichy, K., Bizikova, L., Isaacs, K., Ghezzi-Kopel, K., and Porciello, J. A. 2020. 'Scoping review of adoption of climate-resilient crops by small-scale producers in low- and middle-income countries'. *Natural Plants*, 6: 1231–1241.

Botswana Television (Btv). 2022. 'Documentary on safflower'. Video. https://fb.watch/ikDVobLpGD/.

Emongor, V. E. 2009. 'The major cations and trace elements in Notwane river, Botswana, and its suitability for irrigation'. *Botswana Journal of Agriculture and Applied Sciences*, 5(1): 63–70.

Emongor, V. E. 2010. 'Safflower (*Carthamus tinctorius* L.) the underutilized and neglected crop: A review'. *Asian Journal of Plant Sciences*, 9(6): 299–306.

Emongor, V. E., Oagile, O., and Kedikanetswe, B. 2013. 'Effects of plant population on growth, development, and oil yield of safflower'. *Journal of Agricultural Science and Technology B*, 3: 321–333.

Emongor, V. E., and Oagile, O. 2017. 'Safflower production'. Retrieved from RUFORUM, September 2022. http://www.ruforum.org/sites/default/files/Safflower%20Manual_BUAN.compressed.pdf.

Emongor, V. E., Orabile, P., Phuduhudu, D., and Oagile, O. 2017. 'Effects of genotype on vegetative growth, yield components and yield, oil content and oil yield of safflower'. *Agricultural Science Research Journal*, 7(12): 381–392.

Emongor, V. E., and Emongor, R. E. 2022. 'Safflower (*Carthamus tinctorius* L.)'. In *Neglected and Underutilized Crops: Smart Future Food*, ed. M. Farooq and H. M. K. Siddique, 683–731. London, UK: Academic Press. https://www.elsevier.com/books/neglected-and-underutilized-crops/farooq/978-0-323-90537-4.

Hufnagel, J., Reckling, M., and Ewert, F. 2020. 'Diverse approaches to crop diversification in agricultural research: A review'. *Agronomy for Sustainable Development*, 40: 14.

IPCC. 2021. 'Fighting climate change: Veolia and the climate change crisis'. Retrieved from Veolia, November 2022. https://www.veolia.co.uk/ecological-transformation/fighting-climate-change?gclid=EAIaIQobChMI2ZeVga-P-wIVjqztCh1vrQy6EAAYASAAEgJ-fuPD_BwE.

Janmohammadi, M. 2015. 'Evaluation of the impact of chemical and biological fertiliser application on agronomical traits of safflower (*Carthamus tinctorius* L.)'. *Proceedings of The Latvian Academy of Sciences*, 69(6): 331–335.

Keheng, B. 2022. 'Abundance and diversity of safflower (*Carthamus tinctorius* L.) arthropods in Botswana'. MSc Thesis. Crop and Soil Sciences Department, Botswana University of Agriculture and Natural Resources (BUAN). https://moodle.buan.ac.bw/handle/13049/554.

Keheng, B., Tiroesele, B., and Emongor, V. E. 2022. 'Abundance and diversity of safflower insect pests in Botswana'. Papers Presented in the RUFORUM AGM Scientific Conference, 12–16 December 2022, Harare, Zimbabwe.

Kereilwe, D., Emongor, V. E., Oagile, O., and Phole, O. 2020. 'Nutritional value of safflower whole seed as animal feed in semi-arid southern African conditions'. *African Crop Science Journal*, 28: 103–115. DOI: https://dx.doi.org/10.4314/acsj.v28i1.8S.

Kereilwe, D., Emongor, V. E., Batlang, U., and Malambane, G. 2022. 'Temperature and duration of exposure on chilling injury of safflower'. Papers Presented in the RUFORUM AGM Scientific Conference, 12–16 December 2022, Harare, Zimbabwe.

Kolanyane, M. O., Moroke, T. S., Tiroesele, B., and Emongor, V. E. 2022. 'Influence of nitrogen and phosphorus on seed yield and oil content of safflower'. Papers Presented in the RUFORUM AGM Scientific Conference, 12–16 December 2022, Harare, Zimbabwe.

Korononeo, M. K. 2023. 'Effects of planting date and genotypes on growth, development, yield, and oil content of safflower under irrigated conditions in semi-arid South-East Botswana'. MSc thesis. Crop and Soil Sciences Department, BUAN, Botswana.

Mahalingam, R., 2015. 'Consideration of combined stress: a crucial paradigm for improving multiple stress tolerance in plants'. In *Combined Stresses in Plants: Physiological, Molecular, and Biochemical Aspects*, ed. R. Mahalingam, 1–25 (Cham, Switzerland: Springer International Publishing). DOI: 10.1007/978-3-319-07899-1_1.

Mani, V., Lee, S., Yeo, Y., and Hahn, B. 2020. 'A metabolic perspective and opportunities in pharmacologically important safflower: Review'. Retrieved from MDPI, August 2022. https://www.mdpi.com/journal/metabolites. DOI: 10.3390/metabo10060253.

Marang, M., Emongor, V. E., and Malambane, G. 2022. 'A review of drought tolerance in safflower'. *International Journal of Plant & Soil Science*, 34(10): 140–149. DOI: 10.9734/IJPSS/2022/v34i1030930.

Moatshe, O. G. 2019. 'Genotype and plant density effects on growth, yield, and oil composition of safflower (*Carthamus tinctorius* L.)'. PhD thesis. Botswana University of Agriculture and Natural Resources, Crop and Soil Sciences Department, 302p.

Moatshe, O. G., Emongor, V. E., Balole, T. V., and Tshwenyane, S. O. 2020a. 'Genotype and plant density effects on oil content and fatty acid composition of safflower'. *African Crop Science Journal*, 28: 83–101. DOI: http://dx.doi.org/10.4314/acsj.v28il.7S.

Moatshe, O. G., Emongor, V. E., Balole, T. V., and Tshwenyane, S. O. 2020b. 'Safflower genotype by plant density on yield and phenological characteristics'. *African Crop Science Journal*, 28: 145–163. DOI: https://dx.doi.org/10.4314/acsj.v28i1.11S.

Moatshe, O. G., Emongor, V. E., and Mashiqa, P. K. 2020c. 'Genotype effect on proximate analysis of safflower as green leafy vegetable'. *Journal of Agricultural Science*, 12(11): 1–8. DOI: https://doi.org/10.5539.

Moatshe, O. G. and Emongor, V. E. 2019. 'Genotype, planting density and seasonal effects on phenological stages of Safflower (*Carthamus tinctorius* L.) in Sebele, Botswana'. *International Journal of Science and Research*, 8(12): 1201–1210. https://www.ijsr.net/archive/v8i12/ART20203114.pdf.

Monyame, P., Tiroesele, B., and Emongor, V. E. 2022. 'Response of safflower insects to nitrogen and phosphorus fertilizer application in Botswana'. Paper Presented in the RUFORUM AGM Scientific Conference, 12–16 December 2022, Harare, Zimbabwe.

Mosupiemang, M., Malambane, G., Mathapa, B. G., and Emongor, V. E. 2022. 'Olesin expression patterns and size of oil bodies as a factor in determining oil content in safflower (*Carthamus tinctorius* L.) genotypes'. *European Journal of Agriculture and Food Sciences*, 4(5): 54–60. DOI: http://dx.doi.org/10.24018/ejfood.2022.4.5.570.

Oarabile, P., Emongor, V. E., Phuduhudu, D., and Oagile, O. 2017. 'Effects of genotype on vegetative growth, yield components and yield, oil content and oil yield of safflower'. *Agricultural Science Research Journal*, 7(12): 381–392. https://www.researchgate.net/publication/333001246.

Phuduhudu, D., Emongor, V. E. Oagile, O., and Oarabile, P. 2018. 'Evaluation of the nutritional potential of safflower meal after oil extraction to be used as livestock feed'. *Scientific Journal of Crop Science*, 7(1): 285–293.

Setshogela, B. P., Emongor, V. E., and Tiroesele, B. 2022. 'Influence of petal harvest time and genotype on petal and seed yield, and dye content of safflower'. Paper presented in the RUFORUM AGM Scientific Conference, 12–16 December 2022, Harare, Zimbabwe.

9

SUSTAINABLE COMMERCIAL PINEAPPLE VALUE CHAIN SYSTEM FOR INCREASED YIELD AND INCOME, AND IMPROVED LIVELIHOOD OF SMALLHOLDER FARMERS IN CENTRAL REGION, GHANA

Festus Annor-Frempong

Introduction

Higher Agricultural Education institutions in Africa have been producing graduates who do not meet industry requirements and producing research outputs that do not address the real needs of their societies and communities. The universities in Africa are therefore being challenged to undergo reforms to make agricultural education more responsive to the needs of students, farming communities, industry, and employers of graduates. With the Community Action Research Platforms, RUFORUM supports university activities aimed at enhancing the impact on smallholder farmers. The CARP strengthens university community outreach, supports value chain development, and builds human resource capacities through experiential learning within universities, technical, vocational education, and training (TVET) institutions in Africa.

The Central Region of Ghana is known for pineapple production, which is a source of livelihood support and a food security commodity for many farm families in the region. The smallholder pineapple farmers are faced with production, post-harvest and marketing challenges, and lack healthy planting materials for expansion of farms to be commercially viable enterprises (Zottorgloh, 2014). On the other hand, the depth of knowledge and advanced technologies generated by the University of Cape Coast are capable of addressing the needs of farmers. There exist protocols and methods for quick and easy multiplication of volumes of clean pineapple planting materials (Acheampong et al., 2015), technology for producing biochar and compost to increase the yield of crops (Frimpong et al., 2016) and appropriate management strategies to control leaf blight disease (van der Puije et al., 2015). These innovations had not been extended to smallholder pineapple farmers in the region (Annor-Frempong, 2013).

DOI: 10.4324/9781003387497-12

Value chain development is at the forefront of agricultural development. It involves addressing all the links from the soil to the consumption of a commodity. The pineapple CARP was conceived to facilitate the development of a sustainable commercial pineapple value chain in the Central Region of Ghana. Activities to produce, market, and deliver a commodity to benefit consumers and actors in the chain were initiated to disseminate pineapple technologies to smallholder farmers. The project is based on the premise that when interactions of key actors in the pineapple value chain are encouraged, they can together address the problems (lack of planting materials, soil, disease, marketing and post-harvest losses) of smallholder farmers. The reduction of the problems would lead to more sustainable production, processing, marketing, and utilisation of pineapple products and other by-products by consumers to generate income and eventually improve the livelihoods of smallholder farmers as shown in Figure 9.1.

The project directly involves smallholder farmers from three pineapple growing areas in Komenda-Edina-Eguafo-Abirem Municipality, Abura-Asebu-Kwamankese and Ekumfi Districts, 17 university students, eight university lecturers and researchers; one each small-scale and large-scale agro-processors; two commercial pineapple producers; four staff of the Ministry of Food and Agriculture;

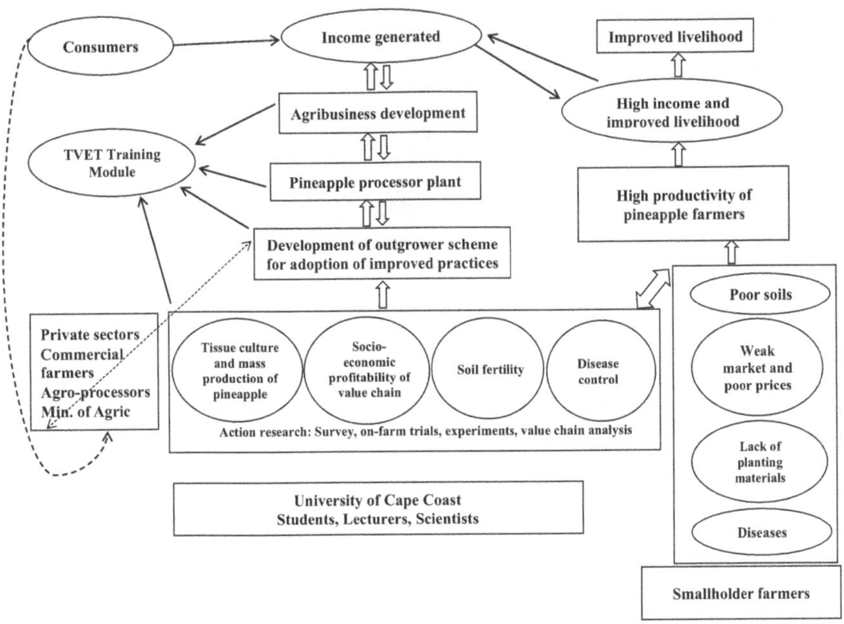

FIGURE 9.1 Conceptual Framework: the development of a sustainable commercial pineapple value chain system with increased yield and improved livelihoods for smallholder farmers in Central Region, Ghana

two NGOs; and students and interns from one agricultural TVET institution and two Agricultural Colleges to develop the pineapple value chain.

Methodology

The project is designed to use stakeholders to apply scientific knowledge to collaboratively address the problems in the pineapple value chain, improve yield, and generate income to sustain the value chain activities. The project, therefore, adopted a participatory action research methodology which is multidisciplinary, interdisciplinary, participatory, action and research oriented. The specific designs that guided students were experimental, mixed methods, and descriptive surveys with the application of unmanned aerial systems (drones), the mobile phone technology system, and computer-assisted technologies. Laboratory and field experiments were used to determine the best varieties of agricultural practices to meet different market needs.

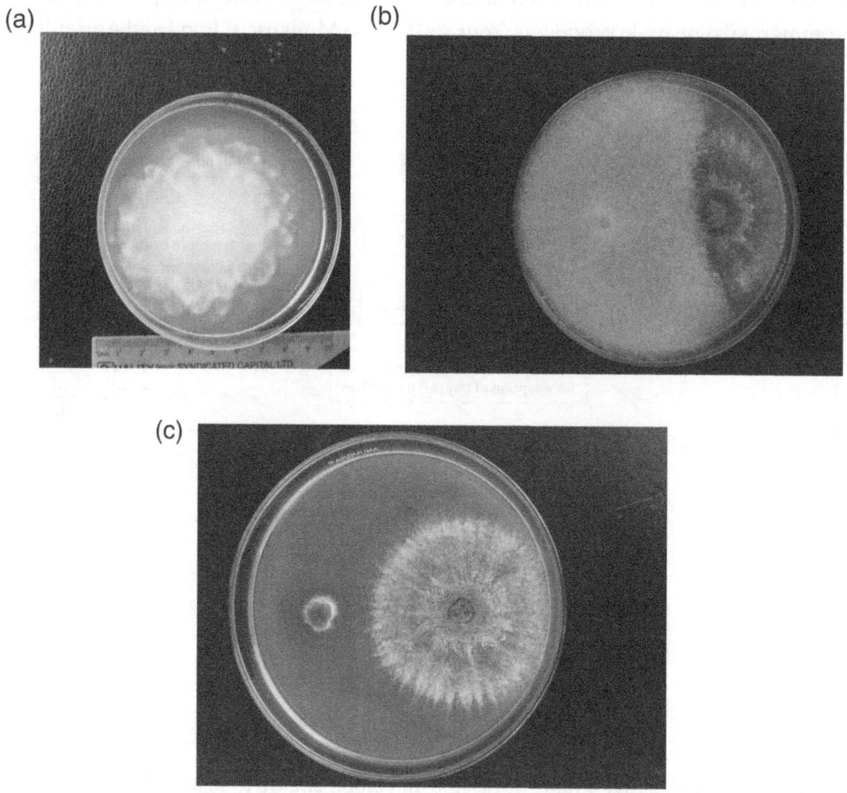

FIGURE 9.2 Laboratory experiments to isolate disease pathogens: (a) colony growth; (b) Impact; (c) effect after tricoderma

FIGURE 9.3a Students preparing split crown multiplication for improved, disease-free planting materials

FIGURE 9.3b CARP team leader with TVET students on field visit

FIGURE 9.4 Planting pineapple split crowns

The project used extension methods such as participant observations, farm and home visits, group meetings and discussions, method demonstrations and field days to engage and train farmers. Computer Assisted Personal Interviews (CAPI) utilised during data collection were a new approach for data collection, analysis, and problem solving in the field (Shaibu, 2021). This included the use of unmanned aerial systems (drones) connected to cellphones and social media. These new approaches also ensured that the project collaborated with public and private sector agencies (pineapple farms and agro-processing companies) to learn commercial production, the establishment of out-grower schemes and marketing of pineapple products. Information systems were established using various media to support the farmers' access to knowledge and markets.

Results

The section presents the results of studies conducted by student researchers, their supervisors, and the principal investigator towards the development of sustainable commercial pineapple value chain for increased yield and income, and improved livelihood of smallholder farmers.

A survey of smallholder farms revealed a higher level of incidence and severity of Pineapple Mealybug Wilt during the pre-flower induction than at the post-flower induction stage (Nyarko, 2019). This CARP research is first to report on the presence of the Pineapple Mealybug Wilt associated virus (PMWaV-1 and PMWaV-3) in Africa (Asare-Bediako and Nyarko, 2019). Mixed viral infections were detected in seven of the 16 PMWaV-infected pineapple samples. Recombination of viral species generated variants with novel genetic features (Asare-Bediako et al., 2020). These results were pivotal in the choosing of the crown as ideal planting material for the split crown multiplication technology. It also informed the strategy developed to manage the viral disease detected on farmers' fields to ensure the sustainability of pineapple production by the project.

Surveys indicated the prevalence of *Phytophethera cinnamomi* and *Phytophethera nicotianae* that cause heart rot disease in the study area. *In vitro* control studies using plant extracts and bio-control organisms (Neem, Prekɛsɛ [*Tetrapleura tetraptera*], Mahogany extract and *Trichoderma asperellum* fungi) showed that all treatments, with the exception of Mahogany, were inhibitory to *P. cinnamomi* with Prekɛsɛ recording the highest mean inhibition index. Application of *Trichoderma* 24 hours before planting gave the highest inhibition percentage (Amoateng, 2021). The implication was that dipping suckers in *Trichoderma* solutions before planting reduces the incidence of *Phytophethera* infections. However, farm sanitation measures were selected to control *Phytophethera* heart rot disease in the 4-acre organic farm established with a farmer group at Nsadwir community (Awuni, 2021) because of the complexity of preparing the solution on a large scale.

Analysis of the coastal savannah acrisol (clay-rich subsoil) revealed low nitrogen, phosphorus, and potassium content. Hanyabui (2020) combined biochar, compost and inorganic NPK fertiliser to significantly increase plant height, number of leaves and fruit weight, length, diameter, and quality of three varieties of pineapple in such soils.

A descriptive survey revealed that most extension agents and farmers had a positive attitude towards commercial pineapple production but a moderate level of knowledge and skills in the production of the crop (Ametepey, 2020). Both had limited knowledge, skills and competencies needed for commercialisation but they perceived it as important, signalling the need for training.

Boakye (2020) used survey design to analyse the value-added activities along the pineapple value chain in selected districts in the Central Region of Ghana and concluded that pineapple production and fruit juice processing are profitable. The fresh fruit marketing business was not profitable due to the very high perishability of the fruits, poor pricing, and unfavourable weather conditions. The findings also showed that farm revenue, capital inputs and cost of planting materials influenced the profitability of smallholder farmers and that of processors depended on the capital cost of pineapple fruits and packaging. The mean efficiency scores of 0.51, 0.45 and 0.13 respectively for pineapple farmers, processors, and marketers imply that they were technically efficient.

Farmer plots were mapped and soil fertility level of pineapple crops analysed using a drone (Shaibu et al., 2019). Shaibu (2021) used concurrent transformative mixed-method design to determine the information needs data from 13 input suppliers, four processors, 30 marketers, 30 consumers, and 233 pineapple producers from KEEA Municipality and Ekumfi District. The active pineapple value actors were agrochemical sellers and smallholder producers. AEAs and market women with smallholder producers expressed a high need for information on inputs, production, and market sources. The information support system (unmanned aerial, mobile phone system, and radio) improved the livelihood of smallholder pineapple producers and enhanced extension delivery.

Lamptey (2019) studied the impact of maturity stage (unmatured, matured, and overmatured), time of harvest (morning, afternoon, and evening) and storage temperature (ambient and refrigeration) of fruits on pineapple juice quality. Harvesting pineapple in the morning followed by ambient storage condition results in consumer-preferred colour and taste of juice. Furthermore, over-matured sugarloaf fruits contain higher total antioxidant and flavonoid content, when harvested in the afternoon and refrigerated. On the other hand, the total antioxidant content of MD2 and smooth cayenne pineapple fruits increases when harvested in the morning and stored under ambient conditions. Pre- and postharvest factors, therefore, influence pineapple juice quality.

Smith (2020) optimised the effects of pineapple juice by examining the yield potential, sensory properties, physiochemical, and microbial quality of microwave and conventional pasteurised juice under room temperature, open-air supermarket, and refrigeration storage conditions. Smith (2020) found significant differences in total soluble solids, titratable acidity, and ascorbic acid among the three varieties of pineapples. The large-sized smooth cayenne variety had the highest fruit juice content. The shelf life of refrigerated microwave pasteurised pineapple juice was 21 days compared to the 14 days for supermarket and room temperature conditions for conventional pasteurised juice. Microwave pasteurised pineapple juice had minimal yeast and mould contamination after three weeks of storage. Smith (2020) used the experiences to develop the *FRUITION CRUSHED* business.

Tokoli (2021) utilised descriptive surveys to examine the influence of attitude, subjective norms, labelling and perceived value of consumer choice towards pineapple fruit juice in the Cape Coast metropolis. Attitude, labelling, and perceived value determined the consumer purchase decision which in turn also influenced the consumers' purchase of pineapple fruit juice.

Tetteh (2020) managed an acre of split crown multiplied planting materials for the commercial production of pineapple with farmers at Nsadwir community whilst Frimpong (2020) introduced farmers at Ankwandah to smooth cayenne variety for commercial pineapple. Awuni (2021) worked with farmers at Nsadwir community to adopt smooth cayenne organic pineapple farming using compost and biochar. Obeng (2021) managed smooth cayenne crowns at the Nyamedom community whilst Watts (2021) managed smooth cayenne ratoon crops with farmers at Ankwandah. Doe (2021) developed tigernut and pineapple products for sale among students at the University of Cape Coast. Asimenu (2022) experimented with propagating healthy planting materials from smooth cayenne corms with Level 400 Agricultural Students from the University of Cape Coast.

Impacts of the pineapple CARP

Increased yield, income, and improved livelihood of smallholder farmers

The project has directly strengthened four farmer groups who have adopted new varieties of pineapple (smooth cayenne and MD2) increasing their incomes. Indirectly, over 10,000 smallholder producers have benefited through training, field visits, radio and television programmes, data collection and a walk-in through which students and resource persons provided information on technologies (Shaibu et al., 2019) and protocols from the pineapple CARP. About 50 acres has been cultivated by the University of Cape Coast as commercial pineapple production over the past four years has been engaging unemployed youth. Thirty agricultural extension agents in the three districts of the project and three cohorts of mid-career extension personnel in the Department of Agricultural Economics and Extension have acquired competence in commercial pineapple production (conventional and organic) technologies.

Effective graduation of postgraduate students and training of B.Sc and TVET students

The pineapple CARP directly supported 17 students that have all successfully completed their respective programmes on schedule with most pursuing further studies. One has established a profitable pineapple and fruit processing business (UCC, 2020a). The PhD graduate is setting up a business for the Department of Agricultural Economics and Extension to provide extension advisory services using drones and CAPI for digital data collection (UCC, 2020b). The pineapple agro-processing centre is established as a hub to develop and incubate businesses.

TABLE 9.1 Pineapple CARP graduate students and research topics

Name of student	*Title of research*
Shaibu Zikiru PhD	Information system support for pineapple value chain and livelihood of smallholder producers in Komenda-Edina-Eguafo-Abirem Municipality and Ekumfi District, Central Region, Ghana
Enoch Ametepey	Assessment of competencies of agricultural extension agents and smallholder farmers for adoption of commercial pineapple production technologies in the Central Region
Francis Padi Lamptey	Assessment of the impact of some preharvest and postharvest factors on pineapple juice quality and safety
Amma Kissiwaa Amoateng	The study of phytophthora heart rot disease of pineapple in some selected districts in the Central Region
Joseph Nyarko	Prevalence of Mealybug Wilt of Pineapple (MWP) Disease and the associated viruses University of Cape Coast
Emmanuel Hanyabui	Effect of pineapple waste biochar and compost application on pineapple yield and quality in a low nutrient Coastal Savanna Acrisol
Kwaku Boakye	Economic analysis of value added activities along the Pineapple value chain in selected districts in the Central Region, Ghana
Oforiwaah Kuukuwaa Smith	Optimisation of microwave pasteurisation and storage on the physicochemical properties of juice from three pineapple varieties
Collins Tetteh Tokoli	Determinants of consumer purchase decision of pineapple fruit juice in the Cape Coast Metropolis
Eric Odum Tetteh	Nursery and management of split crown multiplied pineapple materials with farmers at Nsadwir community in the Central Region of Ghana

FIGURE 9.5 CARP team in Farmer Fields with Principal Investigator

The pineapple CARP has introduced practical pineapple production into the curricula of three colleges with demonstration fields as field laboratories to train over 400 students and interns from the Asuansi Farm Institute in the Central Region, Kwadaso Agricultural College, and Ejura Agricultural College in Ashanti Region. Relationships and collaboration have been established and strengthened across the education value chain as a result.

The rural communities through the pineapple CARP have forged many linkages with a range of service providers and government policymakers. These linkages are creating an enabling environment for the uptake of commercial pineapple technologies leading to lasting benefits to communities. Innovative information systems have shown their value in reaching the farmers and are being taken up more widely.

Challenges

Natural and economic disasters

The long dry spells followed by excessive rainfall made it very difficult to prepare land on time for production. The COVID-19 pandemic had mixed effects on the pineapple CARP project. The lockdown caused a loss of revenue. Nevertheless, post-lockdown prices of fruits soared because of consumer belief that the bromelain in pineapple could cure or prevent COVID-19. Effects of the pandemic and the Russian–Ukraine war have led to high inflation, limited access, and high costs of inputs. Costs of inputs have increased more than fivefold without a commensurate

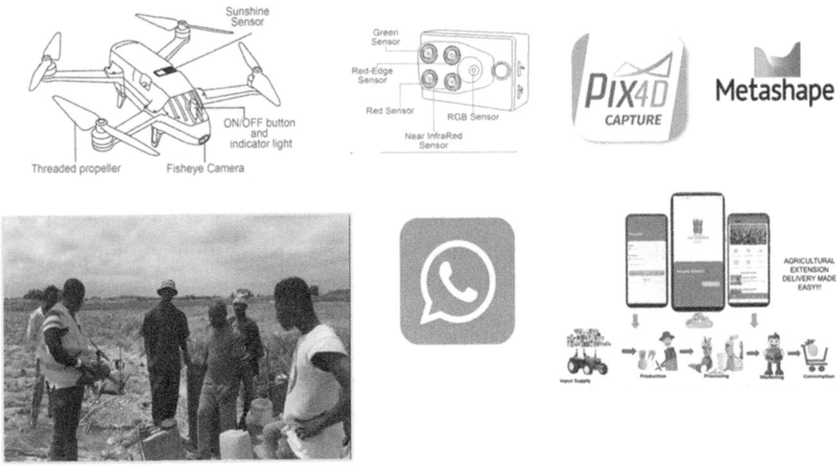

FIGURE 9.6 Information support systems for farmers

increase in fruit prices. These factors have all had impacts on the implementation of the project and the training of students, but the difficulties have been overcome and, albeit with some delays, the students have graduated, and the outcome of the project has been most successful.

Administrative and institutional bottlenecks

Commercial pineapple production is capital intensive. Delays in the release of funds from administrative and procurement bottlenecks in the University make it difficult to follow the cropping cycle. This was experienced when a delay in the release of funds interrupted farmers from receiving tractor services. This is exacerbated when it is necessary to contract services from small enterprise owners that require cash and are not positioned to abide by normal procurement procedures. Furthermore, procurement bottlenecks delayed the purchase of equipment and other supplies for installation. The delays also led to the high cost of project activities because of high inflation.

Facilitating the pineapple value chain development

The pineapple crop takes about 12 to 15 months for fruits to mature for consumption and marketing. Facilitating the linkages among the actors for timely production, marketing, and consumption commercially is an arduous task. Introducing commercial pineapple production technologies to farmers who had not cultivated the smooth cayenne and MD2 pineapple varieties before was difficult. Some farmers were not patient enough to go through the process for 15 months. Others had challenges accessing finance to adopt the new practices, particularly because of the length of the pineapple cycle. There are still information gaps and market gaps to be addressed in future research and engagement.

Conclusion

The CARP approach has helped to identify the problems in the pineapple industry and design appropriate solutions from research focusing on key activities in the value by multiple stakeholders to develop and sustain commercial pineapple production in the Central Region of Ghana. These were achieved notwithstanding the natural and economic disasters, administrative and institutional bottlenecks, and the arduous nature of facilitating the pineapple value chain that acted as barriers to impede progress. The pineapple CARP has increased yield, income, and improved the livelihood of smallholder farmers. There is uptake of improved commercial pineapple production technologies. Graduates involved in the process have gained competencies and enhanced their career progression. The process has also strengthened the relationship between the university and other tertiary agricultural institutions, and with public and private sector agencies in bringing about development. The pineapple CARP has created business opportunities for future students to develop in the production of healthy suckers from crowns as well as biochar, and compost for farmers. The experience from *FRUITION CRUSHED* business set up demonstrates that pineapple juice-based beverages (mixture of pineapple juice with passion fruit, ginger, carrot, or sugar beet) and dry pineapple chips are potential start-ups which could be upscaled. Lessons from the pineapple CARP can be used by the University of Cape Coast to extend the frontiers of research and development and build on this platform to enhance corporate responsibility to communities in the Central Region. Interested readers are referred to the Pineapple CARP University of Cape Coast Blog posts at https://pineapple.ucc.edu.gh/.

Acknowledgements

This platform would not have been possible without the active support of our collaborators in the communities and Asuansi Farm Institute, Kwadaso Agricultural College, Ejura Agricultural College, Ekumfi Fruits and Juices Ltd, Greenfields Farms Ltd, HPW, Ghana Ltd, MacM Farmers Ltd, Marvin Fruits and Petersfield Fruit Processing, the Ministry of Food and Agriculture, Ms. Victoria Abankwa and the supervisors and co-researchers: Dr. William Ghartey, Dr. Selorm Akaba, Dr. Isaac Asante, Dr. Kassimu Issau, Dr. Kofi Atiah, Professor Elvis Asare-Bediako, Professor Ernest Abano, Professor Joseph Kwarteng, Professor Aaron Asare, Professor Ernest Teye, Professor Grace Caselina van der Puije, Professor Kingsley Taah, Professor Kwame Agyei Frimpong, and Professor Samuel Ndzebah Dadzie.

References

Acheampong, S., Galyuon, I. K. A., and Asare, A. T. 2015. 'Effects of sterilization protocols, benzylaminopurine and type of explants on growth initiation of pineapple (Ananas comosus (L.) Merr.) cultures'. *Journal of Basic & Applied Sciences*, 1(3): 50–65.

Ametepey, E. T. K. 2020. 'Assessment of competencies of agricultural extension agents and smallholder farmers for adoption of commercial pineapple production technologies in the Central region'. University of Cape Coast, Ghana. Retrieved from https://ir.ucc.edu.gh/xmlui/handle/123456789/7007.

Amihere, N. 2022. 'Introduction of pineapple production technologies to students at Asuansi Farm Institute'. Dissertation submitted in partial award of B.Sc Extension and Community Development Degree. Department of Agricultural Economics and Extension. University of Cape Coast, Ghana.

Amoateng, A. K. 2021. 'The study of phytophthora heart rot disease of pineapple in some selected districts in the Central Region'. Unpublished MPhil thesis. Department of Crop Science, University of Cape Coast, Ghana.

Annor-Frempong, F. 2013. 'The supervised enterprise projects as a community development tool'. Case Study Series Number 4 on Human Resource Development in Agricultural Extension. Modernizing Extension and Advisory Services (MEAS)/USAID, Chicago and Washington.

Asare-Bediako, E. and Nyarko, J. 2019. 'First report of pineapple mealybug wilt-associated virus-1 and -3 in Ghanaian pineapple'. *New Disease Reports*, 40: 18. http://dx.doi.org/10.5197/j.2044-0588.2019.040.018.

Asare-Bediako, E., Nyarko, J., and van der Puije, G. C. 2020. 'First report of pineapple mealybug wilt associated virus-2 infecting pineapple in Ghana'. *New Disease Reports*, 41: 9. http://dx.doi.org/10.5197/j.2044-0588.2020.041.009.

Asimenu, D. 2022. 'Experimenting with smooth cayenne corms for production of planting materials with level 400 agricultural students of University of Cape Coast'. Dissertation submitted in partial award of B.Sc Extension and Community Development Degree. Department of Agricultural Economics and Extension, University of Cape Coast, Ghana.

Awuni, T. 2021. 'Introduction of organic pineapple farming to Nsadwir farmers'. Dissertation submitted in partial award of B.Sc Extension and Community Development Degree. Department of Agricultural Economics and Extension, University of Cape Coast, Ghana.

Boakye, K. 2020. 'Economic analysis of value added activities along the Pineapple value chain in selected districts in the Central Region, Ghana'. Unpublished MPhil thesis. Department of Agricultural Economics and Extension, University of Cape Coast, Ghana. Retrieved from https://ir.ucc.edu.gh/xmlui/handle/123456789/4804.

Doe, E. 2021. 'Development tigernut and pineapple products'. Dissertation submitted in partial award of B.Sc Agribusiness Degree. Department of Agricultural Economics and Extension, University of Cape Coast, Ghana.

Frimpong, D. 2020. 'Introduction of smooth cayenne pineapple variety and commercial production technologies to farmers at Ankwandah'. Dissertation submitted in partial award of B.Sc Extension and Community Development Degree. Department of Agricultural Economics and Extension, University of Cape Coast, Ghana.

Frimpong, K. A. Amoakwah, E., Osei, B. A., and Arthur, E. 2016. 'Changes in soil chemical properties and lettuce yield response following incorporation of biochar and cow dung to highly weathered acidic soils'. *Journal of Organic Agriculture and Environment*, 4(1): 28–29.

Hanyabui, E. 2020. 'Effect of pineapple waste biochar and compost application on pineapple yield and quality in a low nutrient Coastal Savanna Acrisol'. Unpublished MPhil thesis. Department of Soil Science, University of Cape Coast, Ghana. Retrieved from https://ir.ucc.edu.gh/xmlui/handle/123456789/6521.

Lamptey, F. P. 2019. 'Assessment of the impact of some preharvest and postharvest factors on pineapple juice quality and safety'. Unpublished MPhil thesis. Department of Agricultural Engineering, University of Cape Coast, Ghana. Retrieved from https://ir.ucc.edu.gh/xmlui/handle/123456789/4870.

Nyarko, J. 2019. 'Prevalence of Mealybug Wilt of Pineapple (MWP) disease and the associated viruses'. University of Cape Coast, Ghana. Retrieved from https://ir.ucc.edu.gh/xmlui/handle/123456789/6935.

Obeng, E. 2021. 'Management of smooth cayenne crowns materials for commercial farming'. Dissertation submitted in partial award of B.Sc Extension and Community Development Degree. Department of Agricultural Economics and Extension, University of Cape Coast, Ghana.

Shaibu, Z. 2021. 'Information system support for pineapple value chain and livelihood of smallholder producers in Komenda-Edina-Eguafo-Abirem Municipality and Ekumfi District, Central Region, Ghana'. Unpublished doctoral thesis. Department of Agricultural Economics and Extension, University of Cape Coast, Ghana.

Shaibu, Z., Annor-Frempong. F., and Akaba, S. 2019. 'How drone technology impacts livelihoods: The case of Nsadwir farming community in Ghana'. University of Cape Coast. Retrieved from https://ir.ucc.edu.gh/xmlui/handle/123456789/4319. Please also see https://pineapple.ucc.edu.gh/how-drone-technology-impacts-livelihoods-case-nsadwir-farming-community-ghana.

Smith, O. K. 2020. 'Optimization of microwave pasteurisation and storage on the physicochemical properties of juice from three pineapple varieties'. Unpublished MPhil thesis. Department of Agricultural Engineering, University of Cape Coast, Ghana. Retrieved from https://ir.ucc.edu.gh/xmlui/handle/123456789/6697.

Tetteh, Eric. 2020. 'Nursery and management of split crown multiplied pineapple materials with farmers at Nsadwir community in the Central Region of Ghana'. Dissertation submitted in partial award of B.Sc Extension and Community Development Degree. Department of Agricultural Economics and Extension, University of Cape Coast, Ghana.

Tokoli, C. T. 2021. 'Determinants of consumer purchase decision of pineapple fruit juice in the Cape Coast Metropolis'. Unpublished MBA thesis. Department of Marketing and Supply Chain Management, University of Cape Coast, Ghana.

University of Cape Coast. 30 April 2020a. 'RUFORUM Young Innovators wards inspired me to start my company "Fruition Crushed"'. https://pineapple.ucc.edu.gh/ruforum-young-innovators-wards-inspired-me-start-my-company-fruition-crushed.

University of Cape Coast. 30 April 2020b. 'University of Cape Coast PhD student uses drone technology to improve livelihoods of smallholder farmers'. https://pineapple.ucc.edu.gh/university-cape-coast-phd-student-uses-drone-technology-improve-livelihoods-smallholder-farmers.van der Puije, G. C., Ackah, F. K., and Moses, E. 2015. 'Prevalence of leaf blight disease caused by *Phytophthora colocasiae* in Taro in the Aowin Suaman district of Ghana'. *HortFlora Research Spectrum*, 4(3): 282–284.

Watts, J. 2021. 'Management of smooth cayenne ratoon with farmers at Ankwanda'. Dissertation submitted in partial award of B.Sc Extension and Community Development Degree. Department of Agricultural Economics and Extension, University of Cape Coast, Ghana.

Zottorgloh, T. 2014. *Characterization of smallholder pineapple production systems in Ghana and expert-based perspective on value chain developments*. Wageningen: Wageningen University and Research Farming System Ecology Group.

10

EMPOWERMENT AND POVERTY REDUCTION IN RURAL COASTAL KENYA THROUGH THE CASSAVA VALUE CHAIN

Agnes Mwang'ombe, Samwel Onyango, D. C. Kilalo and P. Wasswa

Introduction

Cassava, *Manihots esculenta*, is a drought-tolerant crop with great potential, which if fully exploited can contribute towards zero poverty and zero hunger. Over 800 million people in Africa produce the crop, which is the third major source of starch globally (Abong et al., 2016). Cassava (*Manihot esculenta*) is a staple crop that contributes significantly to food security and nutrition in Arid and Semi-Arid areas that are prone to drought and other climate change effects. Its productivity is threatened by viral diseases, especially cassava Brown Streak Disease (CBSD), Cassava Mosaic Disease (CMD) and recently cassava bacterial blight. Viruses and bacteria are largely systemic and more often spread in planting material. The practice of farmers exchanging planting materials and the open field production of planting materials has served to increase disease spread. Hence, lack of quality healthy cassava planting materials is a constraint towards the expansion of cassava production and productivity. Markets for cassava roots and its products are not clearly understood. Cassava roots are highly perishable and there is a need to come up with basic techniques to enhance the shelf life and quality of the finished products. The crop is widely consumed and farmer-preferred at the Kenyan Coast in Kilifi County due to its resilience to diverse climatic conditions and relatively high yields of the improved varieties. Production and consumption of cassava was less prevalent in the upland area of Taita-Taveta County where the project also worked.

The University of Nairobi cassava CARP focused on training ten post-graduate students (Table 10.1) by linking them into an innovation platform to address constraints to cassava development. There was close collaboration with a colleague at Makerere University in Uganda. The project built a seed-micro propagation laboratory at the University of Nairobi to clean the diseased available cassava

DOI: 10.4324/9781003387497-13

germplasms, while focusing on farmer-preferred varieties, and rapidly multiply the disease-free cultivars to seed-producing farmers. The project worked with local TVETs to train intern students. The project established working partnerships between the University and the National Government, the local county governments and extension agents, and the farmer CBOs. During the implementation of the project, specifically during the COVID-19 global pandemic, the local farm households in the upland areas came to better appreciate the value of the cassava crop. The cyanide levels in cassava can be a problem and the farmers and processors need to understand the properties of different varieties and the required approaches to processing them correctly. Through the project, the cassava crop increased its potential to become a household security crop. The global pandemic was a blessing in disguise as it highlighted the weak links in the Kenya food system. While farm households had a disrupted supply of vegetables, particularly in the Arid and Semi-Arid Areas (ASAL), cassava was available for utilization. Further, training on the utilization of the crop through value addition diversified the plate options, and farmers enjoyed making cassava-blended meals. The cassava crop proved to be an all-utilized crop from the roots, stems, leaves and peels. The experience

TABLE 10.1 Graduate students in plant pathology, crop breeding, crop protection, animal science, food science and agricultural economics

Name	*Degree*	*Title of research thesis*
Florence Munguti	PhD	Distribution and molecular characterization of viruses and vectors associated with cassava brown streak disease in coastal region of Kenya
Faizo Kasule	M.Sc	Effective isolation distance for prevention of cassava virus spread in farmer-preferred cultivars in Uganda
Patrick Kidasi	M.Sc	Validation of tissue culture and minisett protocols for mass production of quality healthy cassava planting materials
Rose Kamau	M.Sc	Reducing cassava viral pathogen through effective management of whiteflies in Kenyan Coast
Anthony Musera	M.Sc	Prevalence of cassava bacterial blight in Kenyan coast: characterization and early management in planting materials
Elijah Mwangi	M.Sc	Effect of inclusion of sundried and fermented cassava peel meal in broiler chicken diets on performance
Samuel Mwathi	M.Sc	Microbial and chemical profile of fermented cassava leaves from selected Kenyan coastal varieties
S. Ochieng Onyango	M.Sc	Development of protein and mineral enriched cassava root millet-cowpea leaves composite flour from selected popular cassava varieties in Kilifi and Taita-Taveta counties, Kenya
Joe Okoth	M.Sc	Distribution of causal agents of cassava bacterial blight (cbb) in Busia county and determination of resistance to cbb amongst the popularly grown cassava cultivars
Amos Tirra	M.Sc	Economic analysis of smallholder farmers' participation in the cassava marketing value chain in Taita-Taveta and Kilifi counties, Kenya

during the COVID-19 global pandemic raised the economic value of the crop to farm households. Farmers have adopted and shared skills to prepare value-added products both for subsistence and for sale. The cassava CARP sought to empower women and youth in farm households to reduce poverty and improve food security in rural coastal Kenya through a total cassava value chain approach.

Revitalizing the cassava seed system in Kenya through the establishment of a seed incubation lab at the University of Nairobi

The capacity building for micro propagation and certification of cassava planting materials to enhance productivity, incomes and food security and nutrition for smallholder farmers in the Coastal region of Kenya focused on working with women and youth to revitalize the cassava seed system in Kenya. The goal of the project was to improve cassava productivity and reduce the effect of major cassava virus diseases (CMD and CBSD) by availing healthy certified cassava planting materials and developing environmentally friendly viral and bacterial disease management strategies (Kidasi et al., 2021). To achieve the goal the project developed infrastructure (TC lab and greenhouses – for grafting, budding, and hardening seedlings) to support a sustainable, functional, and resilient seed system. The project aimed to produce and avail healthy certified cassava planting materials for farmers to use. Cassava diseases are a major constraint in the productivity of the crop. The viral diseases of CMD and CBSD limit the productivity of cassava (Livoi et al., 2018). The whitefly transmits viral diseases requiring the need for continuous monitoring of the invasive whitefly in cassava for the timely application of management practices. The CARP found that farmers source 83% of their planting materials from own farms or neighbours. They also assessed determinants for farmer-preferred varieties; effective isolation distance for prevention of cassava virus infection; the effect of the use of antibiotics in cleaning the present diseased cassava germplasm; distribution and molecular diversity of whitefly species in Kenya colonizing cassava; and the use of companion cropping as an organic agronomic management practice. Pumpkin was the most important pull crop and coriander was the most effective push crop. Through the sharing of these results and the development of improved access to clean planting material and good practices, the CARP increased farmer yields, and food security.

The cassava diseases spread through the continuous sharing of planting materials among the farmers as one of the routes (Kidasi et al., 2021). The current cassava project sought to determine the farmer sources of cassava planting materials in Kenya to fact find and give recommendations to reduce the spread. The study, which was conducted through four focus group discussions and the use of a semi-structured questionnaire targeting 250 purposively selected cassava farmers in the Kenyan Coastal region in 2018, gave quite impactful insights. The study reported that 83% of the respondents indicated the recycling of cassava cuttings from the previous crops. This is a practice which is traditionally adopted on the Kenya coast.

FIGURE 10.1 Waiting for the rain – a few cassava plants surviving the 2021 drought

While a majority of the farmers acquired seed from their own farm, neighbours (67%) or local markets (5%), the Kenya Agricultural and Livestock Research Organisation (KALRO), as the only formal source of healthy cassava, accounted for 11% of farmers (Kidasi et al., 2021). This reflected the importance of establishing a sustainable healthy cassava planting materials seed system to address the systematic constraints and help develop a viable cassava value chain.

The cassava project trained seed producers at the farm household level. The technology adoption rate varied. An outstanding role was played in raising adoption levels by Amos Mwalaghaya, from Mbale TVET. Amos is one of the farmers who attended the training at the University of Nairobi Kenya. The training was done on the total value chain, clustering the farmers into their segment of interest along the cassava value chain. While a majority of the farmers were interested in cassava production and the sale of cassava roots, Amos focused on seed production. Today, through the support of the project, Amos sources plants from the University of Nairobi for the family farm and uses a screen net. As a result, the county government of Taita-Taveta and the University of Nairobi are engaging to see how best to revolutionize the TVETs curriculum. The TVETs are under the Ministry of Education while our entry point to the community was through the Ministry of Agriculture and so efforts to revise the courses offered at the TVETs received many challenges.

The major viral diseases (CBSD and CMD), cassava bacterial blight caused by *Xanthomonas axonopodis pv. manihotis* and *Xanthonomas axonopodis pv cassava,*

FIGURE 10.2 Tissue cultured cassava seedlings that serve as motherplants, UNO greenhouse

are major constraints in cassava production in Kenya. A study was conducted in the Kenyan coastal regions of Kilifi and Taita-Taveta counties to identify cassava bacterial blight prevalence, distribution, and farmers' knowledge. The study interviewed 250 cassava farmers who were randomly selected. The study reports 61.6% of the respondents identified cassava bacterial blight symptoms in their farms which indicates a wide spread of cassava bacterial blight in the cassava farmer fields in the Kenyan coastal region (Livoi et al., 2021). The study further collected 70 samples randomly. Of the collected cassava samples, 40 were confirmed positive for *X.pvmanihotis*. The study concluded that there is a wide spread of cassava bacterial blight in Kilifi and Taita-Taveta counties thus the need for proper management programs to be deployed in managing the diseases to enhance cassava production (Livoi et al., 2021).

Most of the farmer-preferred (and available) cassava cultivars assessed by the CARP are very diseased with the farmers indicating diverse attributes related to taste and processing for preferring those varieties. Farmer fields were also assessed in eastern Uganda for farmer-attributed preference as well as for the prevalence of CBSD and CMD on those cultivars. The study reports a significant ($P<=0.01$) correlation between cassava cultivars preferred and CBSD severity ($r=-0.56$), CBSD incidence ($r=-0.53$), CMD severity ($r=-0.51$), and CMD incidences ($r=-0.39$). The most preferred farmer cultivar NASE 03 was found to be most susceptible to CBSD and CMD. The results show that CBSD and CMD are prevalent on

FIGURE 10.3 Researchers in the laboratory, UoN (University of Nairobi)

farmer-preferred cassava cultivars and that farmer preference depends on other attributes and not solely disease resistance (Kasule et al., 2020a).

To mitigate the endless spread of diseases through farmer sharing of cassava planting material, the program sought to build a tissue culture (TC) lab for cleaning the existing diseased cassava germplasms which are farmer preferred due to other attributes. The large TC lab at the University of Nairobi has the capacity optimized to produce 5,000 clean cassava platelets weekly. The cleaning procedure capitalizes on the use of antibiotics. While doing the cleaning, a study was conducted to determine the optimal effective combination of antibiotics and the relevant concentration for effective cleaning. Insufficient access to disease-free planting materials is a major constraint in cassava production. Equally, farmers frequently share planting materials over both short and long distances, thus increasing the spread of CBB. The research indicates that the major factor in the transmission of the disease is the recycling of planting materials which harbour latent infections (Livoi et al., 2021). The use of tissue culture techniques at the University of Nairobi tissue culture laboratory has been quite instrumental in the raising of healthy cassava planting materials. The process entails the incorporation of antibiotics in tissue culture media to propagate healthy cassava planting materials. Different concentrations of tetracycline, doxycycline, and streptomycine were incorporated into the media and the plants were monitored for about six weeks at the hardening stage. The seeds are monitored for infection for the period. The results showed that the higher the concentration of the antibiotics, the slower the growth rate of the plantlet. In addition, the absorption/intake rate of antibiotics is better in cuttings (TC) than in rooted plants helping to reduce the antibiotics required. After the production of the cassava plantlets through TC, the plantlets are hardened in a greenhouse or a

FIGURE 10.4 Discussion with farmers' CBO in Kilifi North Sub County

screen house within the University. The mother plants are then grown for about six to eight months until maturity. For rapid multiplication of the plants, the procedure capitalizes on the use of minisett technology to increase plant population. Minisett technology entails the planting of cuttings of three to four nodes.

Given the limited space in the greenhouses, the project grew the cassava in open farmer fields to determine the effective isolation distance for the prevention of cassava virus infection. CBSD and CMD are the major viral diseases of cassava. The study conducted in Uganda used field-sourced cassava planting materials of three cultivars (NASE 03, NASE 14, and NAROCASS 1). It also used TC of the three varieties sourced from the laboratories of the National Crops Resources Research Institute and Makerere University Agricultural Research Institute. Both farmer-supplied (FS) and TC materials were tested at isolation distances of 50, 100, 150, and 250 metres for virus prevention. The 250-metres distance was the most effective for all. The existing recommended distance of 50 metres was ineffective and had the highest infection incidences for CBSD (29.2%) and (CMD 16.1%). The research where there was a high infection rate at the 50-metre distance provided a good opportunity to study the whitefly which is a transmitter of these viral diseases in cassava (Kasule et al., 2020b).

The whitefly, *Bemisia tabaci*, is a crop pest and plant virus vector known to transmit more than 300 plant viruses. The vector transmits viruses that cause two major devastating viral cassava diseases in sub-Saharan Africa and the Kenyan coastal region, namely CMD and CBSD. For effective management of the whiteflies, accurate species identification is vital. The morphological identification approach towards whitefly species limits the capacity to accurately identify new species, especially the presence of cryptic species (Munguti et al., 2022). The cassava project conducted a study using the mitochondrial DNA cytochrome oxidase 1 (mtco1) to sequence four distinct whitefly species namely *Bemisia tabaci*, *Bemisia afer*, *Aleurodicus disperses*, and *Paraleyrodes bondari*, in samples collected from major cassava growing regions in Kenya. The study presents the first report of *P. bondari* (Bondar's nesting whitefly) on cassava in Kenya. We found three *B.tabaci* genetic groups of SSA1, SSA2, and Indian Ocean (IO) putative species colonizing cassava in Kenya (Munguti et al., 2021). The information is useful to inform better management strategies of the whitefly vectors to reduce the impact of cassava viral diseases which continue to be a threat to food security in major cassava-growing regions. It is important to establish continuous monitoring of invasive whitefly species populations in cassava for timely application of management practices to reduce the impact of cassava viral diseases and prevent potential yield losses.

While continuously monitoring whitefly is key to controlling the spread of the viral diseases transmitted, the current project went further to optimize and recommend the most effective push and pull crops to intercrop with cassava to reduce the whiteflies population (Munguti et al., 2018). Companion cropping is a technique that is used in minimizing pests and plant vectors. In Kenya, farmers grow cassava intercropped with legumes, cereals, and vegetables. Pigeon peas, maize, and cowpea are some of the commonly used cassava intercrops. The CARP sought to optimize the use of pullers and pushers in field trials over two planting seasons on the Kenyan coast. The preliminary results of the demonstration plots indicated that pumpkin was a great puller of the whiteflies while coriander was a great pusher. The preliminary findings informed the University of Nairobi of the best push and pull combination to use in the open seed propagation fields.

The project established healthy cassava seed production fields at the University of Nairobi. The project then worked with interested farmers who were trained as seed aggregators. Project partners, Kenya Plant Health Inspectorate Services (KEPHIS), had no documented protocol for certification of stem cuttings. The move triggered a conversation between the concerned stakeholders to work towards the development of a protocol. The University of Nairobi, under the Seed Management Enterprise (SEMIs), is an accredited institution to produce seed. The accreditation has enabled the University of Nairobi to produce healthy cassava seeds and distribute them to the farmers. The demand was such that additional material was also sourced from KALRO Mtwapa (Kenya Agriculture and Livestock Research Organisation) to distribute. The cassava seed system has empowered women and youths at farm households through the distribution of seed and dissemination of the developed technologies for utilization at the farm household level. The project records directly impacting over 4,000 farmers and indirectly reaching out to over 10,000 farmers in open farmer field days and at the agricultural shows.

FIGURE 10.5 Researcher with farmer adopting minisett technology to produce cassava planting material

Value addition and marketing of cassava roots, leaves and peels to minimize post-harvest losses and increase income at farm household in Kilifi and Taita-Taveta counties, Kenya

Food and nutrition security is a major challenge in sub-Saharan Africa. The situation is further escalated in the Arid and Semi-Arid areas. The Kenyan coastal counties, which are predominantly ASAL, experience periodical incidences of hunger. In the region, the cassava crop has been grown but with challenges of diseases and a short shelf life. This requires simple post-harvest and storage protocols to ensure prolonged shelf life (Abong et al., 2016). Cassava value addition ensures maximum utility of the crop and ensures increased availability of food. Equally it raises the economic value of the crop. Simple pre-treatment trials were conducted including blanching, peeling, washing, drying, and fermentation while optimizing for maximum nutrient retention and assessing the cyanide content of different varieties for different uses. The cyanide content of three popular cassava varieties (Kibandameno, Tajirika, Kaleso) on the Kenyan coast ranged 7.8–9.5, 3.4–5.0, and 2.2–2.8 ppb for raw, untreated, and fermented cassava roots (Onyango et al., 2021). Sweet cassava (with low cyanide content) is a rich source of carbohydrates but deficient in other nutrients such as minerals and proteins thus the need to blend cassava with cereals while optimizing for minerals (zinc and iron) and legumes for proteins (Onyango et al., 2020; Owade et al., 2019). Cassava roots are ranked as the third best source of starch after maize and wheat globally.

Nutritious composite flours from cassava roots, millet, and cowpea leaves were formulated, especially for children under the age of five years and expectant women. On the Kenyan coast, pearl millet is also grown. The current project blended cassava roots, millet, and cowpea leaves to give a highly nutritious green porridge rich in carbohydrates meeting the daily requirement of an expectant woman in a single serving. The formulated 20:50:30 flours had protein (8.0%), fat (3.5%), carbohydrates (70%), zinc (0.6 mg100 g^{-1}), iron (0.6 mg100 g-1), and vitamin C (0.3 mg100 g^{-1}) (Onyango et al., 2020).

The cassava crop can be utilized as dried chips, fresh roots, crisps, value-added baked products and equally the leaves as vegetables and the peels as animal feed. Micronutrient malnutrition in Kenya is still high despite the availability of highly nutritious food. While cassava leaves have traditionally been consumed on the Kenyan coast, they are not as widely used elsewhere. Utilization is majorly handicapped due to the high cyanide content. Thus, there is a need for standardized information on cassava fermentation, storage and packaging to propel fast commercialization of the leaves and also to contribute towards the improved availability of micronutrients in diets (Onyango et al., 2018). The CARP sought to optimize the fermentation of cassava leaves. In the study, the upper apical leaves were harvested and grouped into three. The preliminary results confirmed earlier research that a combination of fermentation, dehydration and packaging of the cassava leaves could be optimized to produce dietary supplements that keeps for a long time and can be utilized in many ways (Blanshard et al., 1994).

In Kenya, there is over-reliance on maize as a source of food and feed. Maize yields are relatively low compared to less moisture-reliant crops such as cassava.

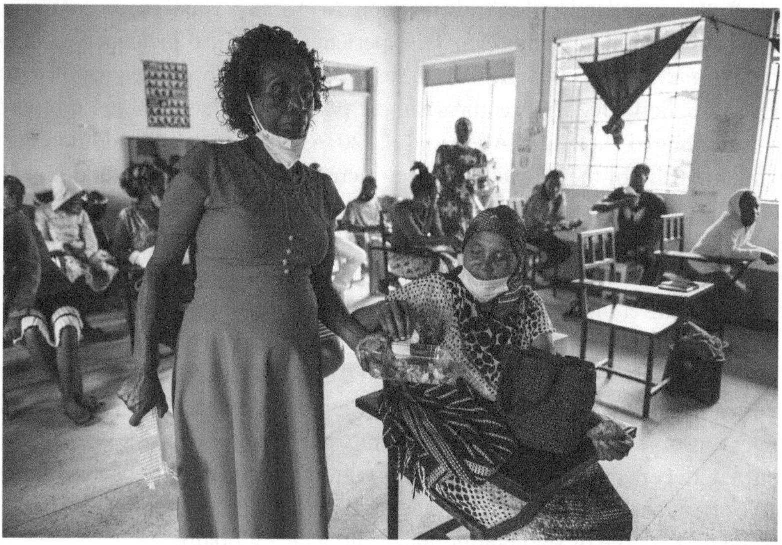

FIGURE 10.6 Workshop on cassava products' preparation, Taita-Taveta

The current conversation is to de-maize in the ASAL areas of Kenya. To achieve that, it is important to increase the economic value of the cassava crop. One way to do this is to add value. While cassava peels have been traditionally allowed to decompose, trials were run on feeding them to chickens. The chickens were fed on different levels of sun-dried and fermented cassava peels. The study collected data on cyanide (HCN) reduction, protein improvement, feed intake, weight gain, and crude mortality rate. Simple sun drying reduced the cyanide in peels to recommended safe-use levels as chicken feed. The chickens fed fermented cassava peels performed significantly ($P<0.05$) better than others in body weight gain and feed conversion efficiency compared to those under the control feed (a locally purchased chicken feed of maize). The cost per kg body weight of the birds fed at the highest level of fermented cassava peels was also lower. To reduce the cost of energy-rich ingredients diets and increase the productivity of chicken, alternative and affordable sources of cassava peels are recommended.

The CARP also documented the determinants of participation in decisions about cassava marketing by smallholder farmers on the Kenya coast. Cassava is an important food crop with high production potential in different ecological zones. In addition to being drought tolerant, cassava is a climate-smart crop promoted for uptake on the Kenyan coast to increase incomes and contribute towards food and nutrition security. During the implementation of the project, farmers were excited to increase production but there was a question of unlocking the cassava markets both locally and for export. The CARP analysed factors that influence a smallholder farmer's decision to participate in cassava marketing in Kilifi and Taita-Taveta counties.

FIGURE 10.7a Farmer harvesting cassava roots in her field

FIGURE 10.7b Farmer feeding dairy with cassava roots

The results of the binary probit model showed that the gender of the head of the house, access to extension services, the price of the cassava products, and the quantity harvested had a significant influence on market participation decisions. Cassava farmers in Taita-Taveta and Kilifi counties participate in the market at different commercialization levels, due to their difference in market orientation (Tirra et al., 2019a). The Household Commercialization Index (HCI) was used to categorize farmers into four different commercial levels. A majority of the farmers explored at least three marketing channels for their cassava (Tirra et al., 2019b). Selling at the farm gate, selling to aggregators, selling to neighbours, and exploring the local markets were some of the highly explored marketing platforms. Credit access, pest management, buying of seed, and the area under cassava all positively influenced the level of commercialization.

The CARP enhanced the technologies and knowledge of the farmers. While they progressively utilized their shared skills, the rate and level of adoption varied from one farmer to the other. Key champion farmers were identified at different segments of the cassava value chain. In Taita-Taveta County, Mama Jane Ndela became a leading cassava value addition expert generating so much income from cassava-baked products that the Taita-Taveta county government adopted her as cassava ambassador changing the perception of her fellow women on the potential of participating in cassava farming. Caroline Dama, a cassava seed farmer in Kilifi County, stated that through the interaction with the project and the subsequent workshops and training through the CARP, she had changed her productivity and increased her acreage under cassava. The farmer groups have also been directly impacted through the farm. They have set up joint farms as demonstration plots while most of those farmers operating as individuals are now growing cassava on at least 0.5 acres. Through the creation of partnerships and collaborations, Renman Africa, a cassava processor based in Malindi, now provides a ready market for cassava roots. The timely partnership with Renman further escalated interest at farm households on their potential to increase cassava production. Some of the farmers have built houses and educated their children from income earned in cassava production since the inception of the project.

Conclusion

The CARP provided practical technologies along the various nodes of the cassava value chain. It established linkages between the university, the farmers and, a wide range of other stakeholders. The project has achieved significant milestones in improved cassava yields, quality, increased profitability, and improved nutrition.

On production, the use of tissue culture techniques provided healthy cassava seedlings through the use of modified tissue culture media. The incorporation of antibiotics and other chemicals in the media eliminated bacteria and viruses giving disease-free cassava tissue culture seedlings. Disease-free cassava tissue culture seedlings were hardened in the greenhouses and provided mother plants which provided plantings at farm fields. Using minisett technology, the cassava mother plants gave rise to several cuttings per plant making it ideal bulking for cassava materials. The use of these technologies has made it possible to produce large quantities of disease-free, certified planting materials for farmers' fields and to provide opportunities for local cassava seed farming. Management of insect vectors has been achieved through intercropping to include crops which serve as repellents or attractants to cassava insect pests such as whiteflies.

The demand for cassava roots has potential but the markets and production areas are not well connected. This project has established farmer-to-farmer marketing of good quality clean planting materials and with the Tissue Culture centre at the University of Nairobi, the continued improvement of farmer access to planting seed is expected to increase. Cassava roots are highly perishable, thus negatively affecting the price paid to farmers. Processing and value addition were embedded into this project to enhance shelf life and profitability of cassava roots. The project has been able to segment cassava roots into various products including a bio-fortified cassava root flour that incorporates cowpea leaf. Nutritious porridge flour, cassava cakes, bread, and chips are some of the differentiated end products which increase returns. Furthermore, some of the cassava farmer groups in both Taita-Taveta and Kilifi Counties in Kenya have set up small businesses around cassava roots, supported through some student start-ups. Cassava leaves have been processed into vegetables with non-detectable cyanide levels and can be included in the farm families' food baskets. Processing cassava roots generate a lot of cassava peels which have been incorporated into broiler chicken feeds giving very impressive growth rates with the acceptability of the broiler meat. Overall, cassava root and leaf processing lowered cyanide levels to acceptable levels for food and feed. Improved processing and product development broadened appeal. This meant that cassava, being a climate-smart crop, is in a better position to become more widespread in Kenya at this time when climate change has become a major challenge to rain-fed agriculture.

The good results obtained under this CARP and the functional cassava value chain hub established at the University of Nairobi have trained students while producing highly relevant research and improving the lives of all the farmers. The students were also trained in agribusiness and have been succeeding in their ventures. Individual farmers, farmer groups, and communities have increased access to nutritious food and are now much more closely linked with service providers and other stakeholders in the value chain. Clean seed protocols and other Kenyan policy impacts have been influenced through this CARP. This collaborative, multidisciplinary project not only linked the university much more closely to farmers and their problems – as well as CBOs and TVETs, local government, extension services, national government, and district research agencies – but it also strengthened links with colleagues in Uganda and other countries in the region. The visibility of the University of Nairobi has been increased and the sustainability of the Hub has been assured by the award of a new grant proposal. The focus for the new project is on the commercial production of disease-free cassava planting materials, cassava root processing, and value addition. This new project is funded by the European Union through FARA, ASARECA and EAFF. The focus countries are Kenya, Rwanda, and Uganda.

Interested readers are referred to the RUFORUM Impact Platform for a film on the CARP, https://ruforum.org/impact/project/university-of-nairobi-cassava-carp/, and to VOICES FROM THE FIELD, Volume 5, October 2020: Transforming Cassava Production through Delivery of Clean Planting Materials and Business Development in Coastal Kenya www.ruforum.org/sites/default/files/VFF%20Vol%20%205%20Web.pdf.

Acknowledgements

This platform would not have been possible without the active support of our collaborators in the communities and Mbale Vocational Training Centre, Dr. Isaac Macharia (KEPHIS) and the Taita-Taveta and Kilifi County Governments and the supervisors and co-researchers: Professor Michael Wandayi, Professor George Abong, Dr. Evans Nyaboga, Professor Willis Oluoch-Kosura, Dr. Hillary Nyang'anga, Professor James Muthomi, Professor Charles K. Gachuiri, Dr. Judith Akinyi Atela, and from Makerere University, Uganda, Professor Settumba Mukasa.

References

Abong, G. O., Shibairo, S., Okoth, M., Lamuka, P., Katama, C., and Ouma, J. 2016. 'Quality and safety characteristics of cassava crisps sold in urban Kenya'. *African Crop Science Journal*, 24: 89–94. DOI: 10.4314/acsj.v24i1.9S. https://www.ajol.info/index.php/acsj/article/view/133106.

Animasaun, D. A., Oyedeji, S., Mustapha, O. T., and Azeez, M. A. 2015. 'Genetic variability study among ten cultivars of cowpea (Vigna unguiculata L. Walp) using morpho-agronomic traits and nutritional composition'. *J. Agric. Sci.*, 10: 119–130. DOI: 10.4038/jas.v10i2.8057. https://jas.sljol.info/articles/10.4038/jas.v10i2.8057.

Blanshard, A. F. J., Dahniya, M. T., Poulter, N. H., and Taylor, A. J. 1994. 'Fermentation of cassava into foofoo: Effect of time and temperature on processing and storage quality'. *J. Sci. Food Agric.*, 66: 485–492. DOI: 10.1002/jsfa.2740660410.

Kasule, F., Wasswa, P., Mukasa, S. B., Okiror, A., and Mwang'ombe, A. W. 2020a. 'Effective isolation distance for prevention of cassava virus infections in Uganda'. *African Crop Science Journal*, 28: 1–13.

Kasule, F., Wasswa, P., Mukasa, S. B., Okiror, A., Nghituwamhata, S. N., Rono, E. C., Mukuze, C., and Mwang'ombe, A. W. 2020b. 'Farmer preference of cassava cultivars in eastern Uganda: A choice beyond disease resistance'. *Agricultural Science*, 2: 169–177.

Kasule, F., Wasswa, P., and Okiror, A. 2021. 'Capacity building for cassava clean seed production using the 250 m isolation distance to address seed challenge in eastern Uganda'. *Discovery Agriculture*, 7: 181–188.

Kasule, F., Wasswa, P., Mukasa, S. B., Okiror, A., Nghituwamhata, S. N., Rono, E. C., Mukuze, C., and Mwang'ombe, A. W. 2022. 'Isolation distance as a phytosanitation option for cassava clean seed in Uganda'. RUFORUM Working Document Series, 19(1): a 27–32. Retrieved from https://repository.ruforum.org/documents/isolation-distance-phytosanitation-option-cassava-clean-seed-multiplication-uganda.

Kidasi, P. C., Chao, D. K., Obudho, O., and Mwang'ombe, A. W. 2021. 'Farmers' sources and varieties of cassava planting materials in coastal Kenya'. *Front. Sustain. Food Syst.*, 6 December 2021. *Sec. Land, Livelihoods and Food Security*, 5. https://doi.org/10.3389/fsufs.2021.611089.

Livoi, Antony, Mwang'ombe, A. W., Nyaboga, E., Kilalo, D., and Obutho, E. 2021. 'Prevalence and distribution of cassava bacterial blight in the Kenyan coast'. *Agricultural Science*, 3(1): 7. DOI: https://doi.org/10.30560/as.v3n1p7.

Livoi, Antony, Mwang'ombe, A. W., Nyaboga, E., Kilalo, D., and Obutho, E. 2018. 'The existence and distribution of cassava bacterial blight in East Africa'. RUFORUM Working Document Series (ISSN 1607–9345), 17(1): 827–834. Available from https://www.ruforum.org/sites/default/files/Livoi%20Antony.pdf.

Munguti, F., Kilalo, D. C., Macharia, I., Nyaboga, E. V., and Mwang'ombe, A. W. 2018. 'Trends in cassava brown streak disease diagnostics for certification of healthy cassava planting Materials and future perspectives'. A review paper published in RUFORUM Working Document Series (ISSN 1607–9345), 17(1): 739–746. Available at https://repository.ruforum.org/system/tdf/Munguti%20F.pdf?file=1&type=node&id=38697&force=.

Munguti, Florence, Dora C. Kilalo, Evans N. Nyaboga, Everlyne N. Wosula, Isaac Macharia, and Agnes W. Mwang'ombe. 2021. 'Distribution and molecular diversity of whitefly species colonizing cassava in Kenya'. *Insects*, Journal by MDPI, 12(10): 875. DOI: https://doi.org/10.3390/insects12100875.

Munguti, Florence Mutave, Evans Nyaega Nyaboga, Dora Chao Kilalo, Hillary Kipkoech Yegon, Isaac Macharia, and Agnes Wakesho Mwang'ombe. 2023. 'Survey of cassava brown streak disease and association of factors influencing its epidemics in smallholder cassava cropping systems of coastal Kenya'. *Front. Sustain. Food Syst.*, 6: 1015315. DOI: 10.3389/fsufs.2022.1015315.

Onyango, S. O., Abong, G. O., Okoth, M. W., Kilalo, D. C., and Mwang'ombe, A.W. 2021. 'Effect of pre-treatment and processing on nutritional composition of cassava roots, millet, and cowpea leaves flours'. *Frontiers in Sustainable Food Systems*, 5. DOI: 10.3389/fsufs.2021.625735. https://www.frontiersin.org/articles/10.3389/fsufs.2021.625735/full.

Onyango, S. O., Abong, G. O., Okoth, M. W., Kilalo, D. C., and Mwang'ombe, A. W. 2020. 'Physico-chemical properties and sensory quality of cassava-cowpea-millet composite flours'. *African Crop Science Journal*, 28(s1). https://www.ajol.info/index.php/acsj/article/view/200157.

Onyango, S. O., Abong, G. O., Okoth, M. W., Kilalo, D. C., and Mwang'ombe, A. W. 2018. 'Post-harvest handling, safety and nutritional contribution of cassava in coastal Kenya'. RUFORUM Working Document Series (ISSN 1607–9345), 17(1): 265–272. Available from https://repository.ruforum.org/system/tdf/Onyango%20Samwel.pdf?file=1&type=node&id=38472&force=.

Owade, J., George, A., Okoth, M., and Mwang'ombe, A. 2019. 'Utilization of traditional processed and preserved cowpea leaves in the Coastal Region of Kenya'. Conference on International Research on Food Security, Natural Resource Management and Rural Development (Kassel), 2017–2020.

Tirra, A. N., Oluoch-Kosura, W., Nyanganga, H., and Mwang'ombe, A. W. 2019a. 'Determinants of participation decision in cassava marketing by smallholder farmers in Taita-Taveta and Kilifi Counties, Kenya'. *Journal of Agricultural Science*, 11(17): 98–109. DOI: 10.5539/ jas.v11n17p98.

Tirra, A. N., Oluoch-Kosura, W., Nyanganga, H., and Mwang'ombe, A. W. 2019b. 'Factors influencing the level of commercialization among smallholder cassava farmers in Taita-Taveta and Kilifi Counties, Kenya'. *African Journal of Agricultural Research*, 14(32): 1584–1592. DOI: 10.5897/AJAR2019.14222.

Tirra, A. N., Oluoch-Kosura, W., Nyanganga, H., Kilalo, D., and Mwang'ombe, A. W. 2018. 'Characterizing cassava marketing value chain in Kenya: A review'. RUFORUM Working Document Series (ISSN 1607–9345), 17(1): 319–327.

11

UNLOCKING OPPORTUNITIES IN THE CASSAVA CARP VALUE CHAIN PROJECT IN NAKURU COUNTY, KENYA

Richard M. S. Mulwa, Joseph W. Matofari, Joshua O. Ogendo, Peter F. Arama and Bockline O. Bebe

Introduction

Nakuru county is one of the high potential agricultural regions of Kenya, contributing 20% to the food basket of the country. However, there are several pockets of arid and semi-arid lands (ASALs) where populations experience frequent maize crop failures but insist on the crop for lack of knowledge on other suitable smallholder crops they can adopt. The overall effect is a compromise on their food and nutrition security and poor incomes. Cassava is a smallholder crop suitable in the marginal regions of Kenya and is a major food crop in the western, coastal and eastern regions. Widespread utilization of cassava is a good option for reducing the demand pressure on cereals in human diets, animal feeds and some industrial products such as beer as well as for enhancing resilience in the face of a changing and variable climate because of its drought tolerance. A national policy initiated in 2007 to foster coordinated cassava value chain development is yet to be finalized and implemented, but the rising demand for food, industrial products and resilience to climate change is fuelling expansion of cassava growing into non-traditional cassava-growing agro-ecological zones. Nakuru County exemplifies such a non-traditional zone where cassava production and agribusiness are being promoted in the ASALs that experience recurrent droughts. The government of Kenya implemented the Kenya Agricultural Productivity and Agribusiness Project (KAPAP) in 2005 that introduced cassava agribusiness promotion in some areas of Nakuru county that experience frequent maize crop failures. Later, in the 2008–2009 severe drought period, promotion of cassava as a drought-tolerant crop was spearheaded by the Mtakatifu Clara Centre in Lare and the KAPAP. This further extended cassava agribusiness promotion in five out of eleven sub-counties of Nakuru County (Njoro, Subukia, Rongai, Naivasha and Nakuru North). In 2010–2015,

DOI: 10.4324/9781003387497-14

a consortium of three public and private institutions (Mtakatifu Clara, Egerton University and Kenya Agricultural and Livestock Organization [KALRO]) partnered to upscale cassava value chain enterprise developments in efforts to contribute to Sustainable Development Goals (SDGs 1–3, 9 and 13) on alleviating poverty and hunger, enhancing good health and well-being, industry and innovation as well as community resilience to climate change. This scaling occurred in the period farmers were searching for alternative adaptable crops to maize which was massively failing from outbreaks of the Maize Lethal Necrosis and recurring droughts. Though the introduced cassava varieties (KME-I, Karembo, Mucericeri, Ndolo, I-96, I-94) had good yields of 10 to 20 tons per hectare, their maturity periods were too long – 18 to 24 months – which meant that farmers needed early-maturing varieties. In stakeholder forums conducted by the three partner institutions, the barriers to further scaling of cassava agribusiness enterprises in Nakuru County were identified to include: (i) long maturity periods of 18–24 months, (ii) the Cassava diseases CMD and CBSD, (iii) limited variety diversity, (iv) consumption limited to fresh cassava and limited product diversity, (iv) poor market linkages and (v) weak farmer institutions to support cassava value chain upgrading.

This cassava CARP project was designed to build on previous efforts to tackle the challenges of curtailing the cassava value chain upgrading in Nakuru County. It aimed at upgrading the cassava value chain through targeted interventions in the ASAL pockets found in three sub-counties of greater Nakuru County. Value chain upgrading is the process of acquiring and transferring technological capabilities and creating market linkages that enable the improvement of competitiveness and trade in higher-value products with product, process and functional upgrading activities. Cassava farmers predominantly grow local landraces that are highly susceptible to CBSD but are still preferred over the improved varieties due to their superior quality attributes such as *in situ* storability, cooking ability, texture of the meal (*ugali*) and taste. The cassava CARP actions focused on selecting sweet (lower cyanide) cassava varieties to breed for CBSD resistance and multiplying them for production and processing into nutritionally enriched food and feed products. Farmer participatory methods were adopted to select preferred genotypes to spur productivity and transform cassava farming from subsistence to a cash economy. The CARP also aimed at revamping cassava processing to open up a market channel for high production. The Subukia Cassava Farmers' Cooperative Society was supported to enhance their aggregation, processing and advocacy capacity.

The initiative aligns with Aspiration 1 of the African Union's Agenda 2063 for inclusive growth and sustainable development as well as Aspiration 6 on people-driven development relying on the potential of women and youth. Locally, the action is contributing to the attainment of food, nutrition and income security through the acquisition and sharing of technological capabilities to encourage wider cassava utilization and the creation of market linkages that enable improvement of competitiveness and trade in high-value products.

Development of the food, industrial and adaptation roles of cassava is currently attracting several targeted interventions with the Kenya Government investing in improving production, marketing and regulations to develop a sustainable cassava industry. This initiative worked with postgraduate students and trained more than 700 farmers and 32 TVET students in cassava cottage product development for increased product ranges and value addition.

The research approach

This project was conceptualized using the sustainable livelihoods model (Ashley and Carney, 1999), which gives focus to the full economic effect of activities undertaken at the lowest societal economic levels. This is a people-centred approach that takes into account the complexity of poor people's lives to promote a more holistic understanding of how they use the full suite of assets and innovations available to them, and the various structures and processes that influence the presence of these assets and innovations.

The project was anchored in fostering innovations in the cassava value chain by promoting the uptake of more adaptable and resilient varieties. The approach also aimed at promoting value addition and the creation of products and services that can attract youth to undertake cassava-based agribusinesses. As a strategy, the project team trained farmers and farmer groups on the different production and processing technologies and also established cassava seed farms in selected farmers' fields to bulk the selected variety(ies), based on resistance to diseases – especially CBSD – and high yields. Considering that most of the value-addition interventions may require capital input, the project purchased affordable processing equipment and exposed the farmers and other stakeholders to their use so that farmer groups could be encouraged to acquire them for their product development.

Baseline survey and designing of action plans

A baseline data collection process was undertaken through a reconnaissance survey and the administration of a household questionnaire. During the reconnaissance survey, key stakeholders (Ministry of Agriculture officials, farmer leaders and cassava welfare group members) in the project sites were identified and geographic locations for the actual survey were mapped. A household questionnaire was administered at the farm level by the project team members and enumerators. A purposive sampling methodology with the aid of farmer leaders in the sites was used to select farmers who grew or had at some time grown cassava. A total of 105 farmers were sampled in the three project sites within Nakuru County. In each of the three sub-counties (Njoro [Lare], Lower Subukia [Waseges] and Rongai [Solai]), 35 farmer households were selected and interviewed. The questionnaire tool captured information on household characteristics including the gender of household heads, level of education of household heads, land tenure, involvement in cassava production, marketing, utilization and any perceived challenges in cassava agribusiness.

The survey findings indicated that up to 90% of households in the survey areas practised farming as their main economic activity while a small percentage, 5%, of households were engaged in non-farm casual work as a primary occupation. Their farm sizes ranged from 0.5 to 3 acres and the average age of the household heads was 60 years, matching the national Kenyan farmers' age of 62 years. The majority of farmers had only primary level education with mostly basic literacy skills: each household had 4–5 direct dependents. Farmers only allocated a maximum of 0.23 acres for cassava as a food security crop with an average production of 2.6 bags per season. Households in the marginal areas of Rongai, Subukia and Njoro were more likely to suffer food insecurity, owing to their dependence on agriculture, small land holdings, low education levels and large numbers of dependents. In all the surveyed areas, farmers indicated that their utilization of cassava for food was limited to boiling and to some extent frying after boiling. There was therefore a need to introduce cassava product diversity in the communities to foster broader usage of the crop for food and nutrition. The areas had no established cassava-based agribusinesses due to the low product volumes from farms. On the issue of cassava production challenges, farmers generally indicated a lack of access to clean planting materials, late maturity of available materials and susceptibility to pests and diseases as the most undesirable variety characteristics. Despite all these enumerated challenges with the cassava value chain, farmers in all the surveyed areas indicated a great enthusiasm to adopt cassava for food and agribusiness if the right support was provided. The project focused on introducing varieties that mature early with improved disease tolerance or resistance, training on good cassava production practices, and the need to provide technologies for the creation of diversified products from cassava to spur production and consumption to enhance food security and nutrition and incomes. The survey findings were applied to design the interventions instituted in the project to address the needs of the farming communities.

Intervention actions

(i) Farmer participatory selection of suitably adapted fast maturing sweet cassava types for the ASALs of Nakuru County

To mitigate the challenge of a limited range of cassava varieties in Nakuru county, farmer participatory activities were conducted in Nakuru, Busia, Migori, Machakos and Makueni counties to enable the collection of popular sweet cassava varieties grown in these counties. Additional improved cultivars were obtained from KALRO centres in Njoro, Alupe and Katumani. In total, 27 varieties of sweet cassava were collected and evaluated in farmers' fields in three multi-locational replicated trials in Subukia, Rongai and Njoro sub-counties to select the most adaptable in these ASAL areas. In each sub-county, the germplasm was participatorily evaluated with farmers from the locality. Participating farmers were trained on the use of the Fukuda et al. (2010) trait descriptor handbook to evaluate the following traits: phenology (time to maturity), root yield, root quality (starch and cyanide content) and CBSD incidence. This helped farmers who had limited experience with cassava to understand the crop deeply. The crop was harvested 12–14 months after planting and sensory evaluations of roots were carried out on the farm to test palatability. Gender perspectives in cultivar selection were taken into account. In each of the three sub-counties, groups of ten representative farmers (five women and five men) were asked to make simple consensus-ranked selections of their five best yielding and safe (not bitter representing lower cyanide) varieties for further bulking and distribution in their localities. The criteria set for yield was a minimum of 30 t ha^{-1}.

Variation in performance was observed as expected due to the different adaptations of the germplasm under testing. This was explained to farmers to be a scientific basis for the selection of varieties for specific sites. Farmers were educated on the fact that not all the tested varieties would do well in their sites and that those that expressed maximum productivity were the most suitable for their particular environments. We informed farmers that although we presented cassava as a widely adaptable crop, some individual varieties are limited in performance because of the crop's sensitivity to genotype by environment (GxE) interactions. Hence to maximize the productivity of particular genotypes, farmers must grow the most adaptable to their environments. Hence, farmers were exposed to knowledge of the fact that to come up with the best-suited cultivars, testing must be carried out across diverse soil and other environmental conditions to be able to select varieties with general and specific adaptabilities. In a feedback session with the farmers, results of the performance of the 27 varieties were presented and Table 11.1 indicates the varieties evaluated.

The engagement with farmers to develop better access to good quality seed are described in Mulwa et al. (2019).

(ii) Development of high-value cassava-based food products

From the gaps identified in cassava utilization, several products were developed, field training on methods of preparation was conducted and the potential for cottage industry uptake assessed. Both baked and deep-fried products were developed. To bring the farming communities into the knowledge of alternative uses for cassava, attributes essential for processing into high-value food products in selected adaptable cassava varieties were determined through laboratory analyses. Analyses were also conducted on the food safety and basic nutritional qualities of the cassava varieties to select those with high dry matter content for further analyses of starch quality. Varieties with low cyanide and high dry matter and starch content were selected for use in product development. In total, 12 products were developed (Table 11.2) out of which the top ranked by farmer groups included fortified cassava-based biscuits, cassava flour cake, cassava grates cake, cassava doughnuts, cassava egg fritters, cassava crisps and cassava chin chin (fried pastry). Assessments of some of the products by home economists and nutritionists pointed to

TABLE 11.1 The 27 cassava varieties tested and their sources in Kenya

Entry	*Variety*	*Source*
1	MH95/0183	Tropical Manioc Selection (TMS), from a farmer in Busia County
2	MM96/0013	Tropical Manioc Selection (TMS), from a farmer in Busia County
3	MM98/0011	Tropical Manioc Selection (TMS), from a farmer in Busia County
4	MM98/3567	Tropical Manioc Selection (TMS), from a farmer in Busia County
5	MM96/7680	Tropical Manioc Selection (TMS), from a farmer in Busia County
6	MM96/2480	Tropical Manioc Selection (TMS), from a farmer in Busia County
7	KME-1	TMS, Kenya Agricultural and Livestock Research Organization (KALRO) Research Center, Njoro
8	MM96/1871	TMS, Kenya Agricultural and Livestock Research Organization (KALRO) Research Center, Njoro
9	MM99/0067	TMS, Kenya Agricultural and Livestock Research Organization (KALRO) Research Center, Njoro
10	MM99/0072	TMS, Kenya Agricultural and Livestock Research Organization (KALRO) Research Center, Njoro
11	Migyera	TMS, Kenya Agricultural and Livestock Research Organization (KALRO) Research Center, Njoro
12	MM99/4884	TMS, Kenya Agricultural and Livestock Research Organization (KALRO) Research Center, Njoro
13	Adhiambo Lera	Local landrace, farmer Migori County
14	Nyar AICAD	Local landrace, farmer Migori County
15	Nyar JICA	Local landrace, farmer Migori County
16	Mabul	Local landrace, farmer Migori County
17	Okonyo Welo	Local landrace, farmer Migori County
18	Madam	Local landrace, farmer Migori County
19	Masisa	Local landrace, farmer Migori County
20	Nyar Maseno	Local landrace, farmer Kisumu County
21	NyaTanga	Local landrace, farmer Migori County
22	Obaro Dak	Local landrace, farmer Migori County
23	Oduwo	Local landrace, farmer Migori County
24	Olomba	Local landrace, farmer Migori County
25	Rao Onyoni	Local landrace, farmer Migori County
26	Selele rabuor	Local landrace, farmer Migori County
27	Karembo	Local landrace, farmer Kilifi County

their suitability for break time snacking among children and could easily replace wheat-based snacks. Enquiring into how the farmers and TVET students would benefit from making these products also identified a lucrative market for these products for school children. These products are displayed in Figure 11.1. Farmer groups and TVET students were given hands-on training on how to make these products as a motivator to spur agribusiness around the products. The recipes for all the products developed are included in a recipe booklet under preparation for publication and distribution to the stakeholders. Other products under development include animal feed supplements – cassava-based animal probiotic pellets and a cassava-based beer. These will be finalized in 2023.

(III) CAPACITY BUILDING

In each of the sub-counties, individual cassava farmers and those organized into registered producer groups were engaged in training on cassava production and upstream value chain issues. Training included visits to the established farmer trial fields where practical assessments were carried out. TVET students at Baraka Agricultural College (BAC) received briefings on the project and were also included as trainees in

TABLE 11.2 Food products developed in the cassava CARP and ranked by participants in order of preference

Rank	*Food product*	*Category*
1	Cassava flour cake	Baked
2	Cassava egg fritters	Fried
3	Cassava grates cake	Wet baked
4	Cassava flour doughnuts	Fried
5	Cassava flour biscuits	Baked
6	Cassava crisps	Fried
7	Cassava chin chin	Baked
8	Cassava flour mandazi	Fried
9	Cassava flour pancake	Fried
10	Cassava flour chapatti	Fried
11	Cassava flour bread sticks	Baked
12	Cassava flour scones	Baked

FIGURE 11.1 Sample of various food products developed in the cassava CARP

cassava production and value-addition trainings. PhD and MSc students were engaged to carry out research in the various aspects of development in the project.

Two PhD students were recruited to work on food product development (Milcah Wambua) and crop improvement (Henry Okwaro) respectively. The PhD student in

(a)

(b)

FIGURE 11.2 The Research Team and partners at the on-farm research sites: (a) CARP team in a farmer's field, Solai; (b) CARP team field discussion in Subukia

TABLE 11.3 Doctoral and master's degree graduates and thesis titles

Name of student	*Thesis title*
Milcah Wambua PhD	Effect of processing method on the properties of cassava flour and cassava biscuits enriched with camel and goat milk
Elizabeth Nderitu	Influence of selected socio-economic and institutional factors on adoption of improved cassava processing technologies among small scale farmers in Migori county, Kenya
Geoffrey Ngeno	Effects of marketing strategies on product line margins among the cassava microenterprises in Migori county
Gad Matayo Sambrumo	Effects of agri-preneurial orientation on growth of cassava based small and medium enterprises in Migori county
Joseph Awuor Odhiambo	Gender disparities in smallholder farmers' access to agricultural productive resources in the cassava value chain in rongo sub county, Migori county
Dimo Caleb	Relationship between selected factors and use of information and communication technology tools among cassava smallholder farmers in Rangwe sub-county
Linda Sitati	Cassava wastes as a source of single cell proteins (scp) for utilization as food and feed supplements using characterized *Trichoderma harzianum*
Richard Maina	Lager Beer Production from Cassava Hydrolysate

food science characterized cassava for food safety and product development quality. Work on the assessment of cyanide content in leaves and roots was also undertaken to establish their vegetable quality and safety. Four master's students, two in community studies and two in agribusiness, were trained. Two additional master's students in food science worked on the development of cassava probiotic products for animal feed supplementation and a cassava beer, respectively.

(iv) Knowledge sharing on cassava and cassava-based products

At the onset, the project endeavoured to enlighten the farming communities and other stakeholders on the benefits of growing and utilizing cassava for basic food and processing into high-value products. Farmer stakeholders' meetings and consultations were organized where project aims and outcomes were shared and recommendations for quality standards for cassava-based products were discussed. The project was also serialized twice on national media bringing to the attention of the country the impact of cassava production and value addition on the transformation of the food security situations among dwellers of ASAL pockets in the high potential areas of the country. Further stakeholder engagement at the policy level will involve the County Executive Committee (CEC), Members of Agriculture, Marketing, Trade and Industry, county government portfolios and relevant national ministries, the Kenya Plant Health Inspectorate Services (KEPHIS),

Kenya Agricultural and Livestock Research Organization (KALRO), Horticultural Development Authority (HDA), Kenya Bureau of Standard (KEBS) and the private sector – East African Breweries and animal feed millers. The aim is to revive discussions on the national cassava policy document and push for the full implementation of the Kenya Food and Nutritional Policy which directs the use of 15% blending of cassava flour in maize meal.

Project outcomes and impact pathways

(a) Personal scientists' level

At a personal level, the CARP initiative was an eye opener to the scientists who were, prior to this, largely laboratory-based scientists. The CARP project brought us out of labs and placed us in a real-life environment having to communicate and transfer our technological knowhow with the people at whom these technologies were targeted. Many lessons were learnt including the skills required in community mobilization and having to address pre-conceived negative views that cassava is dangerous to eat. This project introduced cassava, a new crop in the target areas, to a community that was not familiar with its production or consumption. There were understandable fears about the crop. Cassava had been associated with serious food poisoning episodes for a long time and the communities in Lare, Solai and Subukia were aware of this. These fears were allayed by educating the communities on the two major types of cassava (sweet and bitter) and training them on how to identify, process and use them. The scientists in the project have gained valuable experience and skills in the implementation of community-based projects.

(b) Institutional level

Hosting of the cassava CARP project at Egerton University helped to increase the visibility and opening up of the University to communities beyond the neighbourhoods of its location. This project enabled the University to extend other services to farmers from Subukia and Rongai sub-counties besides their interactions on the cassava value chain. Two farmers in lower Subukia were linked to the University and assisted to improve their aquaculture operations by accessing fish fingerlings and advice from the University's Agro-based Science Park. The University's Animal Science Department, acting through the project, reached out to participating farmers starting new dairy operations.

The cassava CARP project directly benefited Egerton University through the purchase of small laboratory equipment for students' research and documentation. Through the project, a motorized cassava chipper, baking ovens, a documentation camera and plant tissue culture equipment were acquired. Additionally, the project helped to raise the graduate student portfolio of the University. One PhD and two MSc students have been trained under direct research support from the project apart from another five MSc students who conducted research in the cassava value chain were sponsored through the larger TAGDev programme. Additionally, four undergraduate students were trained on the tissue culture of cassava as part of their final-year research projects. Of the four, one (Bernard Ngeno) joined a master's programme at the Hebrew University of Technology to pursue research in the tissue culture of root crops. Another student from that cohort (Bernard Sichangi) is currently pursuing an MSc in horticulture and conducting his research in cassava tissue culture with the aim of developing a protocol for *in vitro* multiplication and virus-free seed material production of one of the varieties selected by farmers in the CARP project. All MSc students whose research was aligned to the cassava value chain have published research papers in refereed journals as a condition for their graduation. This has raised the publications portfolio of the University. Five research papers have so far been published (Nderitu et al., 2019; Wambua et al., 2020; Ngenoh et al., 2020; Awuor et al., 2021; Sambrumo et al., 2022) and three more are under review.

As a result of engaging in the cassava CARP project, our team also competed in the national climate-smart agriculture programme call that was targeting the unlocking of cassava value addition potential for economic growth in the country. The team won a grant of Ksh. 23 million with Egerton University as the lead entity for the project in three ASAL counties of Kenya.

(c) Community level

The cassava CARP project was designed to address some real felt needs of the target communities to deal with food insecurities and income generation. This project reached out to over 700 farmers who were directly impacted and more than 4,000 indirectly. Farmers participated in setting up the variety evaluation trials, hosted the trials and visited the trials as their learning centres to have peer learning. Additionally, farmers in the three target sub-counties attended nine training sessions on the production and value-addition/product development of cassava. A quick impact of these interactions is that seven individual farmers who participated in the training are already bulking cassava varieties for the sale of planting materials and two farmers are making and then selling various cottage cassava food products. One cassava farmers' cooperative society was facilitated to revamp processing activities through the installation of their processing plant. This cooperative had hit a hitch in the installation of equipment donated to them by the World Bank but did not have the facilitation to install it. Overall, the cassava CARP project has enabled the improvement of household food security for farmers that adopted cassava production, even on small land holdings.

(d) National level

The cassava CARP project developed 11 food products that were mainstreamed among participating farmers to diversify their diets. A presentation of these products at the national food systems forum resulted in six of them being included in the climate-smart foods panel under the Kenya Climate Smart Agriculture Programme (KCSAP). These six products have been adopted as TIMPs for national food and nutrition initiatives under the programme. Farmers in two KCSAP counties (Kisumu and Busia) have had two value-addition training sessions on these products by the cassava CARP project team.

Challenges

Several challenges were faced as the cassava CARP project was conducted. These ranged from logistical, health and societal challenges which the project had to surmount to be successful. The major ones are enumerated below:

a) *Logistics*: There were infrastructure challenges at the beginning of the project. Roads leading to the selected sites of operation were in a deplorable state. The project team at one point had to leave one vehicle in the field and await rains to subside for two days before it was retrieved. Hired transport providers at some point were reluctant to provide transport due to the high costs of repairs they would incur on providing services over the short distances covered. It should be noted that the project had a rate of payment per kilometre covered. However, after eight months of project implementation, a major tarmac ring road connecting Solai and Subukia was constructed which shortened the time of travel between the two sites from two and a half hours to 40 minutes. Also, the road connecting Stoo Mbili on the Egerton-Narok highway and Elementaita on the Nakuru-Nairobi highway was upgraded to tarmac grade in the second year of the project, making access to the Lare site more convenient.
b) *Health*: The COVID-19 pandemic negatively impacted the delivery of the project. Egerton University was closed twice (first for eight months and second for two months) due to flare-ups of the pandemic in Kenya with severely restricted access to the targeted project areas. On the resumption of University operations, there were heavy access restrictions to the Solai and Subukia sites as the disease had taken root there. Finally, when the pandemic eased, the project team had to plan contacts with farmers involving the local administrations as the protocols had to be observed strictly. On a sad note, the project lost two participating farmers to the pandemic. The pandemic also affected students' activities leading to delays in completing their studies.
c) *Society*: A project syndrome developed with the delivery of the cassava CARP project. The targeted areas for the project had not received extension services for a long time. Thus, on the first contact with the farmers, there were many expectations beyond the scope of the project. Additionally, a positive challenge arose as the project drew to a close. There is a high demand for cassava planting materials arising out of the interactions between the team and the farmers. The bulking sites established are not sufficiently serving the available demand and

the selected farmers run out of materials. This is being addressed through the CARP by linking the needs of the farmers to the county government so they can budget for the crop as a value chain of concern.

d) *Industry*: The industry partner's (East African Breweries Limited) expectations vis-à-vis small-scale farmer capacities has become a challenge. Despite the fact that the industry has a large market for cassava, the low produce volumes and capacities of the farmers cannot as yet feed into the requirements of EABL for their scale of production. An assessment of the requirements for cassava to fill in the demand at the brewery company indicates that a national campaign on production is required to sustain supplies to this particular industrial market.

Conclusion

In this project, a community participatory approach linking research and production with the cottage industry has been tested and is being adopted at the national level for strong inclusive business growth among women, youth and rural cassava producers. The model is aimed at creating interdependence between production and utilization of raw materials for domestic and targeted basic industrial uses. New dimensions of cassava for feed and brewery applications focuses on opening up new product demand pathways. East Africa Breweries Limited (EABL), a private brewery, is exploring the possibility of substituting barley with cassava in its brewing, hence increasing sustained demand for cassava. Cassava is also a cheap substitute for cereals in feed production and can add to the sustained demand for cassava. Capacity building in a variety of development and adaptation, and good agricultural and entrepreneurial skills provided to cassava value chain actors will enhance the sustainability of cassava enterprises.

Interested readers are referred to the Egerton University cassava CARP film on the RUFORUM Impact Platform https://ruforum.org/impact/project/egerton-university-seed-cassava-carp-2021/.

References

Ashley, C., and Carney, D. 1999. 'Sustainable livelihoods: Lessons from early experience'. DFID, London.

Awuor, J. O., Mulwa, R. M. S., and Openda, N. O. 2021. 'Gender disparity in cassava farmers' access to agricultural productive resources in Rongo Sub County, Migori County, Kenya'. *African Journal of Agricultural Research*, 17(9): 1161–1171.

Fukuda, W. M. G., Guevara, C. L., Kawuki, R., and Ferguson, M. E. 2010. 'Selected morphological and agronomic descriptors for the characterization of cassava'. International Institute of Tropical Agriculture (IITA), Ibadan, Nigeria. 19 pp.

Mulwa, R. M., Arama, P. F., Nderitu, E., Ng'eno, G., and Oduor, S. 2019. 'Voices from the Field: Enhancing cassava seed bulking and production skills in Nakuru County'. Vol. 2. Egerton University. RUFORUM accessed March 2023. https://repository.ruforum.org/documents/voices-field-enhancing-potato-and-cassava-value-chains-performance.

Nderitu, E. M., Kamuru, S. M., and Matofari, J. W. 2019. 'Influence of institutional factors on adoption of improved cassava processing technologies among small scale farmers in Migori County, Kenya'. *African Journal of Pure and Applied Science Research*, 1(1): 26–32.

Ngenoh, G. C., Kariuki, I. M., Gathungu, E. W., and Kiprop, S. K. 2020. 'Factors influencing the choice of marketing strategies among cassava microenterprises in Kenya'. *African Crop Science Journal*, 28(s1): 117–129.

Sambrumo, G. M., Ngigi, M. W., and Mutai, B. K. 2022. 'Influence of agripreneurial orientation constructs on growth of cassava-based small and medium enterprises in Migori County, Kenya'. *Asian Research Journal of Agriculture*, 15(4): 45–55.

Wambua, M., Mulwa, R. M. S., Arama, P. F. Atieno, S. A. and Ogendo, J. O. 2020. 'Evaluation of popular cassava varieties for yield and cyanide content under ASAL conditions in Kenya'. *African Crop Science Journal*, 28(s1): 71–82.

12

SEED POTATO PRODUCTION IN NAKURU, KENYA

Outcomes and implications of an active multi-stakeholder platform

Anthony Mwangi Kibe, Mariam Mwangi and Agnes Oywaya Nkurumwa

Introduction

The potato (*Solanum tuberosum* L.) is the second most important food and cash crop in Kenya after maize and the third globally, after rice and wheat. It is an important source of income for approximately 800,000 small-scale farmers in Kenya, employs 3.3 million people along the value chain (KEPHIS, 2016) and contributes over KES 50 billion to the Kenyan economy. About 161,000 ha is cultivated, producing 2 million metric tonnes annually (FAOSTAT, 2019). In 2013, the average potato productivity in Kenya was 14.4 MT/ha (Andrew et al., 2021a). In 2018 however, the productivity dropped to about 5 t/ha only. Potato yield is largely dependent on the quality of the seed. Low production and access to quality seed potato remains a major challenge among smallholder farmers in the County (Ong'ayo et al., 2020). Farmers often recycle seeds beyond the recommended number of times, thereby encouraging disease and pest build-up. Additionally, the centralized and oligopolistic nature of the potato seed industry in Kenya leads to high transactional costs and pricing of certified seed. This places a burden on resource-constrained farmers preventing them from accessing certified seed (Okello et al., 2017). Other challenges include the dilution of certified seed or sale of fake seed and weak phytosanitary control and traceability at border points of ware potato that end up being used as seed, thus introducing more diseases and pests. Smallholder farmers have little training and extension about the importance of using clean seed or about seed potato management at the farm level. According to Ong'ayo et al. (2020) farmers who interacted more frequently with knowledge change agents benefited from information on clean seed potato multiplication.

Good quality seed potatoes are in short supply and are expensive and not easily accessible to smallholder farmers resulting in poor yields, low incomes and

DOI: 10.4324/9781003387497-15

spreading disease by continuously reusing seed from their ware potato crops. Nakuru County, where this CARP is based, is the highest producer of seed potato, accounting for 70% of the 6,700 tons produced nationally. This is below the estimated demand of 30,000 tons (KEPHIS, 2021). With the exception of ADC Molo which is a government parastatal, other national and county governments, research organizations and universities do not have sufficient funding for research and development in the seed potato sector. This CARP showed that small farmers cannot easily participate in the formal seed chain which requires a complex process of registering and paying for inspections. The use of poor-quality seed is a major factor leading to low tuber yields (Demo et al., 2016; KEPHIS, 2016). Nearly 95% of smallholder potato farmers (SHF) use degenerated seed due to inadequate supply and inaccessibility to quality seed in the country (McGuireet, 2016).

The seed systems in Kenya are generally categorized as formal (i.e., certified at 2–3% of total seed potato production in Kenya); semi-formal (i.e., *clean seed* at 2–3%: this is equivalent to quality declared seed (QDS) in Uganda and Rwanda which is not regulated to certification by KEPHIS in Kenya) and informal *farmers' seed* (95%). Tissue culture, with the careful production of minitubers is used to produce early generated seed (EGS) which is much less prone to disease. Kenyan-produced certified seed potato is mostly produced by the public sector, that is, Agricultural Development Corporation (ADC) Molo and the Kenya Agricultural and Livestock Research Organization (KALRO). There are 30 registered seed merchants nationally: both public and private with Nakuru having the highest number (i.e., 11) of seed merchants. ADC Molo still remains the largest public institution producing certified seed potatoes at 2,625 MT per year. The Kenyan-bred potato varieties are only 22 among the 67 potato varieties in the National variety list (NPCK Variety Catalogue, 2021). More than 40 Dutch varieties have been introduced and adopted by seed companies as have others including French, Irish and Scottish varieties. Only five Kenyan varieties are commercially produced and much of the seed potato is imported making it unaffordable to small farmers. This problem has been exacerbated by international seed potato suppliers that are supported to bring in their own seed potato propagules, some of which are potential carriers of diseases and pests. These are safer in Europe where they do not increase in incidence in their temperate climates but which have no natural enemies and or winter seasons in the tropics to suppress their negative effects in Kenyan conditions. Kibe articulates the development of the seed-potato CARP to address these issues in more detail in Voices from the Field (Kibe et al., 2019).

After the CARP inception, a presidential taskforce was initiated to prioritize potato as one of the flagship crops to achieve their food security goal and Kenya's Vision 2030. Egerton University, Njoro, ADC Molo, the National Potato Council of Kenya (NPCK), Baraka Agricultural College, the Nakuru Smallholder Farmers Association (NASFA) and other partners have been collaborating on this CARP and have supported the taskforce. Working with smallholders in Nakuru County it has become increasingly apparent that a program to transform the seed potato sector to include smallholder farmers is required and must be sustained to enhance the productive potential of potatoes for effective contribution to national and international goals.

This CARP project engaged in original research to address access to high-quality seed potato by smallholder farmers by promoting the use of available innovations as well as inform interventions along the entire value chain. The research involved the participation of students, staff of Egerton University (EU) and Baraka Agricultural College (BAC), a TVET working closely with farmers in Molo. The project directly strengthened the capacity of approximately 5,800 smallholder farmers through the platform that has now evolved into the Nakuru Potato Corporative Union (NPCU) with the support of Nakuru County government and the Ministry of Agriculture Livestock and Fisheries (MoAL&F).

The inability of farmers to adapt to, or mitigate against, weather variability and climate change has resulted in variability in farm productivity and total production of potatoes in Kenya and Africa. In addition to poor seed, declining productivity is affected by fungal disease, especially in heavy rainfall years. The productivity drop to about 5 t/ha only in 2018 was attributed to a high incidence of late blight enhanced by low temperatures (<10°C in July) associated with high relative humidity (> 90%) and continuous heavy seasonal rainfall. Cultivation of poor-quality seed material susceptible to fungal diseases, inappropriate knowledge platforms to provide early warning and advisory services on suitable mitigating or coping strategies and lack of access to funds to purchase agrochemicals in a timely manner resulted in lowering of yields in farmers' fields and thus total production. The SP-CARP through the multi-stakeholder platform approach has brought seed potato actors to understand the causes of low yields and helped to concertedly solve common seed production problems (Kibe and Ngumba, 2020). This has resulted in increasing access to high quality seed potato of over 10% of smallholder farmers in Molo Sub-County (Mutinda, 2019).

Multi-stakeholder platforms to support active engagement

The design of the CARP as a platform has supported the building of relationships and triggered engagement of the university at County level to help to address the problems identified. The platform was initially established with research and extension agencies operating in Nakuru County and with Baraka Agricultural College (BAC). The platform evolved considerably during the course of the project and catalyzed the integration of a wide range of stakeholders (Kibe and Ngumba, 2020).

On-farm field days on seed potato production were initially carried out at Egerton University Tatton farm and at Baraka College, before then setting up in on-farm sites in five areas.[1] The CARP project deliberately involved the farmers as co-researchers with the students. It established close liaison with Baraka to undertake a larger scaling for the supply of good quality (but not necessarily certified) seed and to train more farmers across Nakuru to scale out the findings from the farmer-student research in the CARP. Egerton University in collaboration with the national potato council and fifteen other stakeholders, conducted a field day at Baraka Agricultural College that had over 1,100 participants in 2019. In June 2022, a field day on a large farm in Mau Narok included smallholders and had over 2,000 participants.

Taking this SP-CARP forward, the Kenya Climate Smart Agricultural Project has added other Kenyan universities, national and country potato cooperatives, large private sector input suppliers, potato processors and distributors to the platform. From 2018 to date, the SP-CARP platform events have promoted potato and seed potato development activities in Nakuru, Nyandarua, Elgeyo Marakwet and Nyeri Counties with participants at each gathering ranging from 40 to over 1,100. The platform has also triggered other field days supported by private service providers and NGOs. Collaboration amongst the multi-stakeholder platform has helped to provide better links and business associations of producers with markets. The CARP, including the government agencies ADC Molo and AGRICO, organized outreach to promote their research and product outputs in collaboration with service providers. This initiative linked farmers with a wide range of service providers, such as Quipbank, which leases farm machinery and transport vehicles. Farm Machinery Distributors, CASE and Toyota Tsusho demonstrated mechanized land preparation and harvesting. The Cereal Growers Association (CGA) along with their (youth) extension service provider, the Farm to Market Alliance, partnered with agrochemical and fertilizer input providers to support and promote mechanized spraying and management of ware potato production. Syngenta Foundation and Sunculture have supported field days including demonstrations of solar pumps for irrigation of French beans (a rotation crop) for Frigokan, a fresh produce exporter which in turn partially funds the construction of farm ponds. Knowledge providers assisted the CARP and helped to link farmers to markets, including Tropical Heat, Njoro Canning, the Department of Defence, a food processing firm in Gilgil, Sereni Fries and potato aggregators such as Twiga Foods. This extensive collaboration has also evolved into effective engagement for addressing policy issues culminating in a draft Potato Strategy 2021–26 that has been submitted to the legislature for County approval, but the political process moves slowly and requires ongoing facilitation to maintain impetus.

The university deepened the farmer field research by undertaking specialized laboratory and greenhouse activities at the university to support improved potato seed and ware potato management. Guided by Faculty members students developed proposals, conducted university and on-farm research and disseminated results through organized stakeholder workshops, field days, internship workplaces and publications. To maintain communication and joint activities with researchers, farmers, students, associated farmer groups, knowledge providers and agro-enterprises, WhatsApp groups were established. This has subsequently developed into a very effective online platform for planning meetings, executing outreach activities and the sharing of output (results) and experiences. The students were closely engaged in all these activities and strengthened links between the university, from the tissue culture laboratory and greenhouse trials to the farm trials and the research outcomes and sharing.

Results and discussion: enhanced production skills and access to quality seed potato

The students engaged have worked with farmers, gained experience and improved their commitment to rural communities and enhanced their personal skills and employability. This multi-stakeholder platform has increased access to high-quality seed potato. The impact of this seed increment in the last two years is yet to be evaluated. The SP-CARP has helped to catalyse more research in this under-researched important crop. Including students from other departments made facilitating the research more difficult but resulted in a more effective impact. Work done has increased the number of seed producers in Nakuru, with Egerton and BAC being

TABLE 12.1 Seed potato postgraduate CARP students

Name	*Program of study*	*Thesis title*
Redempter Mutinda	MSc. Agri-Enterprise Dev.	Factors influencing agripreneurs' participation and investments in clean seed potato enterprises in Molo, Nakuru county, Kenya
Sheku Gbollie	MSc. Soil Science	Potassium and Sodium Extraction in Cocopeat for Potato (*Solanum tuberosam* L.): minituber production under varied calcium nitrate soaking durations and cocopeat-pumice media
Andrew Waaswa	MSc. Agricultural Extension	Relationship between selected factors and the practice of climate-smart agriculture among smallholder potato farmers in Gilgil sub-county, Kenya
Jerusa Ong'ayo	MSc. Agri-Enterprise Dev.	Networking capability, adoption tendencies and commercialization: case of decentralized clean seed potato multiplication agri-enterprises in Nakuru county, Kenya
Satognon Felix	MSc. Soil Science	Effects of Nitrogen and Irrigation Regimes on Performance of Potato (*Solanum tuberosum* L.) Grown in Mollic Andosols
Mutuku Kilonzi Jackson	MSc. Crop Protection	Efficacy of *Trichoderma asperellum* seed treatment and ridomil application in the management of late blight in potato
Betty Thuo	MSc. Gender & Dev.	Gender Disparities in Adoption of Climate Smart Agriculture Practices amongst Seed Potato Farmers in Olkalau, Nyandarua County
Felistus Makau	MSc. Horticulture	Effect of nutrient and soilless media on growth and survival of apical rooted potato varieties

FIGURE 12.1 Students with farmers preparing seed beds

registered as seed producers by KEPHIS in 2019, marketing over 50 tons and 70 tons respectively in two years. These entities directly reach small farmers with their good quality seed and are expected to raise not only the ware potato productivity in Nakuru but also to train and develop small farmers into seed potato growers and to enhance the distribution networks. This SP-CARP has resulted in triggering additional resources and the Seed Potato Kenya Climate Smart Agricultural Project (SP-KCSAP) will scale the platform activities to Nyandarua, Nyeri and Elgeyo Markwet Counties.

Action Research Outputs

Improved research and practical skills of Faculty and students of Egerton University and Baraka Agricultural College through action research within the seed potato value chain

Public agricultural extension service providers are usually graduate products of universities and TVET institutions. Many in the past have graduated with very little practical exposure to the agriculture industry. The SP-CARP approach of providing experiential learning through active engagement in value chain platforms has enabled the students and Faculty staff to get much-needed skills and build confidence. Many graduates are now actively engaged in outreach and commercial efforts in four potato-growing Counties in Kenya as well as in their home countries in Africa. Multi-stakeholder engagements spearheaded by the CARP revealed that the low levels of seed potato production in the dominant potato-growing areas of Nakuru, namely Molo and Njoro sub-Counties, were attributed to a number of factors as summarized from students' and Faculty publications.

Social economic factors that affect seed potato production and agri-enterprise development

Mutinda a SP-CARP student, noted a decline in the amount invested in seed potato enterprise from KES 167,248 (US$1 = 100 KES) in 2017 to KES 82,961 in 2019. This was attributed to a decline in productivity in farmers' fields, thus increasing seed scarcity and the costs of seed and, ultimately, lowering production (Mutinda et al., 2020). The dominant seed system used by the agri-preneurs was own farm-saved seed (60%), whose productivity is low due to a decrease in quality resulting from recycling seed. A majority (89%) of seed potato agri-preneurs were selling uncertified seed with the remaining 11% were selling certified seed (Mutinda, 2019). This was much higher than the national average of 2.6% in 2014 implying that farmers are adopting the use of certified seed in Nakuru County. Factors that positively and significantly ($P< 0.1$) influenced agribusiness investments in *clean seed potato* were age, sex (male), access to a seed store, better selling price, acquired training, longer family history in the seed potato business, more years spent in school and years in potato farming (Mutinda, 2019), education level, farmer group membership and distance from where the seed was sourced (Amwine et al., 2019). Furthermore, farmers that had cultivated more area under seed potato had an annual household income greater than US $6,000, frequently got extension service and had initially started with a higher acreage under production ($P< .01$) which significantly increased the level of investment in *clean seed potato* enterprises (Mutinda et al., 2020). Subsequent studies by Ong'ayo et al. (2020) and Andrew et al. (2021b) (CARP students) confirmed that partner knowledge, ownership of transport and storage facilities and frequent interaction with certified seed potato producers influenced investments positively. Awareness and adaptation of climate-smart agricultural practices (CSAPs) that enhanced seed production in Nakuru County differed significantly between men and women as well as the communication information pathways used. Amongst other CSAPs investigated by Andrew et al. (2021c), smallholders were aware of irrigation but had low uptake due to costs and lack of technical skills. The use of potato seedlings and minitubers for planting early generation seed (EGS) was both at low awareness and practice. Women were more aware of available CSAPs than men through neighbours and friends, but men had a greater access to media, mobile phones and extension staff information communication pathways than women. Mainstreaming of CSAPs information was recommended for achieving higher adoption.

Research results underpinning impact

1. *Early generation seed production in laboratory and screen houses*

Tissue culture techniques' have become useful technologies for producing disease and pest-free seed potato and rapid multiplication of rooted apical cuttings generated from either plantlets or microtubers. As part of the CARP project Makau et al. (2022) recommended the use of *Murashige and Skoog* media mixed with 40 gL^{-1} sucrose and 0.5 mgL^{-1} gibberellic acid to enhance the growth of tissue-cultured explants of *wanjiku, unica* and *shangi* varieties. This *in vitro* treatment gave the highest plant survival of 90% and enhanced shoot length to a maximum of 10.3 cm for the Wanjiku variety. Growing rooted apical cuttings after hardening them in the greenhouse is a challenge to many early generation seed (EGS) producers, resulting in very few smallholders adopting the production of EGS.

At the EGS level, plants are still delicate and require protective screen houses and skilled use of growing media, i.e., hydroponic systems that do not harbour quarantined diseases and pests. Cocopeat, a by-product of the Coconut (*Cocos nucifera* L.), is an important (mostly imported but with great local production potential in coastal Kenya) soilless media, that contains high levels of potassium (K), sodium (Na) and electrical conductivity (EC). Methods for extracting these elements, thus lowering EC are yet to be standardized. Gbolle et al. (2021), another CARP student, reported that the washing of cocopeat returned the EC and pH values to within suitable ranges for potato growth, irrespective of $Ca(NO3)_2$ application rates and soaking durations. Effective supplementation of Ca and N and optimal reduction of K and Na by 78.44% and 92%, respectively, could be achieved with 100 g of $Ca(NO3)_2$ 1.5 kg^{-1} of cocopeat in 15 litres of water under a soaking duration of 36 hours. This standardization is important for use in the hydroponics industry in Kenya as it saves time and money and it also enhances the productivity of seed. There were strong negative correlations between Ca and Na, Ca and K and between Na and EC. A strong positive correlation was observed between Ca and N and between Ca and EC. Therefore, highly saline/alkaline soilless media can hinder the uptake of essential nutrients and should be avoided. These results have informed the seed potato investments and production at both Egerton and Baraka Agricultural College which are producing certified seed.

2. *Agronomic technical innovations and management practices*

Satognon et al. (2021a) reported that for optimizing nitrogen use efficiency (NUE) of apical rooted cuttings of potato in a mollic Andosol, farmers should apply 90 kg N ha^{-1} with an irrigation regime of 85% of the crop evapotranspiration (ETc). However, for the highest seed potato yield production, an irrigation regime of 100% ETc and 130 kg Nitrogen ha^{-1} can be applied in three split applications at 10 (40%), 30 (40%) and 50 (20%) days after planting Satognon et al. (2021b). The cumulative actual crop evapotranspiration (ETa) estimated was on average 201.4, 302.1,

342.4 and 402.8 mm under 50%, 75%, 85% and 100% ETa respectively. Potato plant height, number of branches per plant, tuber number per plant and DM were responded to N application rates, while total tuber yield, marketable tuber and HI were more responsive to the interaction of both factors than a single factor.

Integrated disease management that incorporates biological agents provides sustainable opportunities in managing late blight for enhanced tuber yield production. Latently infected seed tubers (particularly in informal systems accounting for 95% of seed in the country) initiate late blight epidemics early in the cropping season. The blight forces farmers to adopt a shorter spray-interval regime, thus increasing the costs of production. Kilonzi et al. (2021) opined that the combination of *Trichoderma asperellum* at 66% concentration with a 14-day spray interval provided better late blight management than conventional application of fungicides. Results suggested that seed treatment by dipping using 66% *T. asperellum* suspension could increase the fungicide application interval by seven days while improving on yield. This approach reverses the risks of latently infected yet symptomless seed tubers that are the common source of primary *P. Infestans* inoculum responsible for late blight epidemics early in the cropping season.

3. *Marketing and market access*

Many small-scale farmers are unable to get marketing contracts to supply potatoes because of stringent quality and quantity requirements demanded by product

FIGURE 12.2 Identifying clean seed

FIGURE 12.3 Weeding on-farm plot

uptake companies. This requires them to improve their potato grading and identify appropriate markets. Collective marketing has been identified as one of the best strategies to improve the participation of small-scale farmers in accessing markets. Nyasula, a CARP student, noted that only 7.6% of small-scale farmers were involved in collective marketing in four wards of Nakuru County (Nysaula et al., 2019). Women, youth and farmers with high literacy levels were high amongst members who participated in collective marketing. Lack of access to better markets, lack of storage facilities and lack of training on collective marketing, as well as a feeling of being exploited by buyers and lack of trust among group members reduced participation in collective marketing. Farmers able to marshal capital of over US$6,000 annually to sustainably grow their seed potato enterprise (Mutinda et al., 2020) prefer to go it alone rather than work within platforms that tend to have restrictive operational rules and interdependency. The CARP worked to help strengthen farmer groups to assist them to aggregate using social media. However, Starlite cooperative, Portech and Mau Narok Sacco are smallholder farmer groups that have managed to hold on together to support their members in accessing seed potato markets. The Nakuru Farmers Association and Mogun SHF have had management issues due to apathy and ageing of their members. Groups that have a higher percentage of youthful leadership in their 30s are motivated and better workers when given support. Means to support nurturing mothers and youths to participate in workshops and field days need to be given priority.

Impact on livelihoods

The participatory approach led to strong uptake by the farmers and the spread of ideas across farmer groups in Nakuru. Participating farmers significantly increased their incomes. As a result of their engagement in the research, they produced and used better-quality seed and improved agricultural practices. This was especially evident for the participants that produced good quality seed for themselves and to

FIGURE 12.4 Exhibiting products at Baraka Agricultural College Show

FIGURE 12.5 Mbaira family in front of new home built from seed potato profits

sell locally on the informal market. A number of the farmers significantly improved their living conditions as a result of the project, and they have also become much more active in supporting farmer groups to support potato growing.

BOX 12.1 THE EXPERIENCE OF A SMALLHOLDER FARMER WITH THE SEED POTATO CARP

For Daniel Kamiti, chairman of the Starlight Farmers Cooperative Society, all the different facets of the CARP were starting to come together to make a difference in the lives of people in the community. Like so many areas in Nakuru County, his community was deeply affected by the post-election violence of 2007 and 2008. Hundreds of people lost their homes and livelihoods in the violence, as their homes, fields and food stores were burned. Daniel and many others fled to the nearby forests to take shelter when violence erupted and then later to IDP camps. Even nearby Baraka College was temporarily converted into an IDP camp. The community here first started growing potatoes around 2012 working with the NGO WorldVision. The crop was seen as a means of bringing the community together after the post-election rupture. This land has seen deep patterns of change and disruption over generations.

Sixty years ago this land was being farmed by white settlers, and the remnants of old houses and infrastructure remain, including an old broken concrete water reservoir standing on the top of the hill. When Kenya gained independence and the settlers left, tensions simmered in the community over who had a right to the land after several generations of occupation. Today, Daniel grows potatoes interspersed with cabbages on a downward slope, on land once farmed by his father and before that his grandfather. The land has also been subdivided over the generations. Today his share amounts to a tiny fraction of the 117 acres once farmed by his grandfather. After the death of his father in 1983, he depended on farming for his livelihood.

On a downward slope, Daniel grows potatoes intercropped with maize and beans. The rows of food crops are interspersed with patches of grass, left as barriers to retain water on the slope. Through the CARP, Daniel has harnessed the scientific expertise of the university, as well as the physical resources of seeds and inputs, in combination with his own intimate knowledge of his land, developed over so many years of farming here. His potatoes are ready to harvest now, but with a smile, he says he intends to store them in the ground a while longer. This is because potatoes are now selling for KSH 1200 a sack but in a couple of months' time, the price will double. Building a livelihood as a potato farmer has been a slow and steady process but it is now starting to pay off. After losing most of his crop in the initial seasons, he is now able to produce his own seed potato. He now has a secure livelihood and is able to comfortably pay school fees and invest in new inputs such as a dairy cow and a second-hand vehicle with which he now can deliver his produce to market. 'Even if I had another option, I wouldn't do anything else besides farming. It is a life of independence,' he says.

FIGURE 12.6 Storing seed potatoes

Sustainability: improve governance and performance of the seed potato value chain in Nakuru County

Despite all the challenges that evolve in developing a working potato multi-stakeholder platform, it is evident that the highly unstructured and thus fragmented seed potato industry has benefited from the organizational structure that is built around formal collaborative arrangements. It is important to continue the engagement to avoid duplicated efforts. The MoAL&F, KALRO and agriculture faculties in universities largely work in '*silos*.' This has been partially addressed by the SP-CARP and the climate-smart program which have brought organizations together in their efforts to support the seed potato sector. This requires ongoing investment to strengthen collaboration.

An outcome of the CARP was the Nakuru County Stakeholder Platform including government agencies, universities, the private sector and farmers, established in 2020 under the auspices of a national agricultural project.[2] This, in turn, resulted in the formation of the Nakuru Potato Corporative Union (NPCU). The success of developing a vibrant potato platform is very much dependent on the networking capabilities of individuals working in host institutions. Similarly, a successful seed potato agri-preneur also depends on his or her networking approach (Kibe and Ngumba, 2020). According to Ong'ayo et al. (2020), networking is a necessary entrepreneurial capability that can ease the development and utilization of linkages to gain access to resources, knowledge and complementary assets at minimal transaction costs, hence, improving enterprise performance. The sustainability of the CARP activities is evident in the mushrooming of similar ongoing platform members' collaborative activities of validating the best innovations on-farm. In particular, the mobilization of stakeholders for public participation in the gazetting and enforcement efforts of the Crops (Irish Potato) 2019 Regulations was effectively done through the WhatsApp platform. Viazi-soko, an online mobile application, developed by NPCK will continue providing useful information on sourcing and marketing seed potatoes in specific agro-ecological zones. In combination with knowledge outputs from action research, the Faculty members and students have developed training and extension materials for disseminating relevant seed and climate-smart technical and management innovations to farmers and the agro-industry. Currently, demonstrations are being conducted on-farm in five counties with support from the World Bank-funded Kenya Climate Smart Agricultural Project (KCSAP).

Conclusion

Infrastructure to support EGS using hydroponics and further multiplication on-farm is now in place at Egerton University and BAC. Egerton Seed Unit is recognized as a seed potato producer by KEPHIS. The SP-CARP has supported postgraduate students and Egerton is now receiving many applications from prospective students interested in pursuing seed potato studies. The profile of Egerton University as a source of information, training and quality seed potato production has expanded in the country and region. The livelihoods of farmers in the area have improved: they are benefiting from the university. Egerton University, BAC, National and County Governments and donors should therefore seek ways to enable the community action research program (CARP) platform approach to scale out. The policy of enabling the public agriculture departments and parastatals to co-develop agriculture products and exhibit them during outreach events showed the policy to be a good approach that should also be scaled. This will help the ministry have the motivation to spearhead the organization and host the Nakuru Potato Stakeholders platform's activities. The government agriculture extension officers should be enjoined with Egerton University and KALRO in the research and development agendas of the County. This will help the line ministries to be motivated to take charge of their mandate to spearhead the organizing and hosting of the Seed and Ware Potato Stakeholders platforms' activities.

Interested readers are referred to the Seed Potato CARP on the RUFORUM Impact Platform: https://ruforum.org/impact/project/egerton-university-seed-potatoes-carp/.

Notes

1 At Likia (Mau Narok), Turi, Elburgon (Molo), Kuresoi North (BAC) and Kasambara (Naivasha).

2 The World Bank's National Agricultural Rural Inclusive Project (NARIG-P) funded seed potato farmers to federate their activities across the country.

References

Amwine, L., Nkurumwa, A. O., and Mulu-Mutuku, M. W. 2019. 'Socio-Economic Factors that Influence Adoption of Clean Seed Potato (*Solanum tuberosum*) by Small Scale Farmers in Njoro, Kenya'. *Society of Educational Research and Evaluation in Kenya*, 1(1). https://www.researchgate.net/publication/347086737.

Felista, Mutheu Makau, Mariam Mwangi, Maurice Edward Oyoo, Anthony Mwangi Kibe, and Judy Oggema. 2022. 'Effects of Sucrose and Gibberellic Acidon Growth and Survival of Local Potato (*Solanum tuberosum* L.) Varieties in vitro in Kenya'. *European Journal of Biology and Biotechnology*, 3(4). http://dx.doi.org/10.24018/ejbio.2022.3.4.372.

Gbollie, S. N., Mwonga, S. M., and Kibe, A. M. August 26, 2021. 'Effects of Calcium Nitrate Levels and Soaking Durations on Cocopeat Nutrient Content'. *Journal of Agricultural Chemistry and Environment*, No. 10: 372–388. https://doi.org/10.4236/jacen.2021.103024.

KEPHIS. 2016. 'Kenya Official Variety Register'. Kenya Plant Health Inspectorate Service, Nairobi.

KEPHIS. 2021. 'Kenya Official Variety Register'. Kenya Plant Health Inspectorate Service, Nairobi.

Kibe, A. M., Nyasulu, L., Aarakit, P., and Njihia, S. 2019. 'Voices from the Field: Enhancing Potato and Cassava Value Chains Performance'. Vol. 2. Egerton University. Retrieved from RUFORUM, March 2023, https://repository.ruforum.org/documents/voices-field-enhancing-potato-and-cassava-value-chains-performance.

Kibe, Antony Mwangi and Rahab W. Ngumba. December 2020. 'Establishing and Scaling the Seed Potato Community Action Research Multi-Stakeholder Platforms: Lessons & Experiences from Kenya'. *International Journal of Innovative Research and Knowledge*, 5(12).

Mutinda Redempter. 2019. 'Factors Influencing Agripreneurs' Participation and Investments in Clean Seed Potato Enterprises in Molo, Nakuru County, Kenya'. MSc. Agri-entreprenureship thesis. Egerton University, Nakuru, Kenya.

Mutinda Redempter, Edith Wambui Gathungu, Dennis Kyalo Wambua, and Antony Mwangi Kibe. 2020. 'Factors Influencing Agripreneurs' Participation and Investment Levels in Clean Seed Potato Enterprises in Nakuru County, Kenya'. *African Crop Science Journal*, 28(Issue Supplement). DOI: 10.4314/acsj.v28i1.12S.

Mutuku Kilonzi Jackson, Mafurah Juma Joseph, Nyongesa Moses Wabomba, and Antony Mwangi Kibe. 2021. 'Efficacy of *Trichoderma asperellum* Seed Treatment and Ridomil® Application in Managing Late Blight on Potato'. *World Journal of Agricultural Research*, 9(2): 42–52. http://pubs.sciepub.com/wjar/9/2/1.

NPCK Variety Catalogue. 2021. https://npck.org/wp-content/uploads/2021/10/NPCK-2021-CATALOGUE-17.pdf.

Nyasulu, Lucky, Mulu-Mutuku, M. W., and Nkurumwa, A. O. 2019. 'Participation of Small-Scale Farmers in Collective Marketing of Potato in Molo Sub-County, Nakuru County, Kenya'. *The Authentic Knowledge: African Journal of Pure and Applied Science Research*, 1(1).https://serek.or.ke/conferencejournals/index.php/AJPASR/article/view/24.

Okello, J. J., Zhou, Y., Kwikiriza, N., Sylvester Ogutu, Ian Barker, Elmar Schulte-Geldermann, Elly Atieno, and Justin Taj Ahmed. 2017. 'Productivity and Food Security Effects of Using of Certified Seed Potato: The Case of Kenya's Potato Farmers'. *Agriculture & Food Security*, 6: Article# 25. https://doi.org/10.1186/s40066-017-0101-0.

Ong'ayo, M. J., Gido, E. O., Ingasia, A. O., Mwangi, M., and Kibe, A. M. 2020. 'Role of Networking Capability, Socio-economic and Institutional Characteristics on Adoption

Tendencies of *Cleanseed* Potato Agri-Enterprises in Central Rift Valley, Kenya'. *African Crop Science Journal*, 28(s1): 1–9.

Satognon, Felix, Seth F. O. Owido, and Joyce J. Lelei. 2021a. 'Water Use Efficiency and Nitrogen Use Efficiency of Apical Rooted Cuttings of Potato Grown in a Mollic Andosol'.*Environmental Systems Research*, 10(1). DOI: 10.1186/s40068-021-00242-4.

Satognon, Felix, Seth F. O. Owido, and Joyce J. Lelei. September 2021b. 'Performance of Apical Rooted Cuttings of Potato Grown in Mollic Andosols under Different Nitrogen Fertilization and Irrigation Regimes'. *Heliyon*, 7(9): e07999. https://www.sciencedirect.com/science/article/pii/S2405844021021022.

Waaswa, Andrew, Agnes Oywaya Nkurumwa, Anthony Mwangi Kibe, and Joel Ng'eno Kipkemoi. 2021a. 'Climate-smart Agriculture and Potato Production in Kenya: Review of the Determinants of Practice'. *Climate and Development*, 14(1): 75–90. DOI: https://doi.org/10.1080/17565529.2021.1885336.

Waaswa, Andrew, Agnes Oywaya Nkurumwa, Anthony Mwangi Kibe, and Joel Ng'eno Kipkemoi. 2021b. 'Communicating Climate Change Adaptation Strategies: Climate-smart Agriculture Information Dissemination Pathways among Smallholder Potato Farmers in Gilgil Sub-County, Kenya'. *Heliyon*, 7(8). DOI: https://doi.org/10.1016/j.heliyon.2021.e07873.

Waaswa, Andrew, Agnes Oywaya Nkurumwa, Anthony Mwangi Kibe, and Joel Ng'eno Kipkemoi. 21 September 2021c. 'Understanding the Socio-economic Determinants of Adoption of Climate-smart Agricultural Practices among Smallholder Potato Farmers in Gilgil Sub-County, Kenya'. *Discover Sustainability*, 2(41). https://link.springer.com/article/10.1007/s43621-021-00050-x.

13

INCLUSION OF FARMER COMMUNITIES IN VALUE CHAIN DEVELOPMENT

A case of potato in South-western Uganda

Johnny Mugisha, Peter Wasswa, Napoléon Heri Bahati-Kajunju and Abel Atukwase

Introduction

Potato (*Solanum tuberosum* L.) is ranked the third world's most important tuber crop grown by small-scale farmers for cash and food (Muthoni et al., 2013). In sub-Saharan Africa, production has more than doubled since 1994 with about 70% of the growth concentrated in eastern Africa (Lutaladio et al., 2010). In Uganda, its production is concentrated in the highland areas; the South-western highlands of Kabale and Kisoro produce about 60% of total output. Recognising the increasing importance of potato, the International Year of the Potato encouraged policymakers, agronomists, and economists to re-evaluate the potato in its role as an ideal crop to sustain food security through crisis (FAO, 2010). The FAO, therefore, recommended that the major potato-growing countries should have a seed policy guide and regulations that address the issues of seed potato in order to promote a well-coordinated and functional potato value chain. To address these policy issues, the government of Uganda developed a Potato Framework Implementation Plan (FIP) which, among others, focuses on supporting production, distribution, and marketing of seed and ware potato and provision of an enabling policy framework.

In Uganda, the potato has become a staple in the diets of many households. The crop matures within 3–4 months and in some areas, it is grown up to three times a year. This makes it a good pathway for enhancing the incomes of households, especially the low-resource households (Mugisha et al., 2017). However, potato productivity has stagnated below potential due to a multitude of factors. Very low yield estimated at 7.5 t/ha-1 has been realised compared to 20 t/ha-1 that is reportedly achievable on research stations (Arinaitwe et al., 2017). The potato value chain is characterised by, among others, limited access to quality seed (Mbowa and Mwesigye, 2016).

DOI: 10.4324/9781003387497-16

Two MSc students worked on this challenge of limited access to quality seed and efficient utilisation of the available seed. Scarcity of other farm resources, especially land, also poses a challenge. In the main potato growing areas, arable land is very small, yet farmers grow a wide range of crops which compete for the land. In such a situation, it became crucial to improve land use efficiency. However, most research on potato in Uganda had looked at the crop as though it was only grown as a monocrop and in isolation from other crops, yet in reality many farmers intercropped potato with other crops. This implied that the research had not benefited those farmers, partly explaining why the adoption of some research recommendations had remained low. One MSc student used a combined experimental and modelling approach to develop an optimised cropping system.

Potato has high energy output per unit land, high carbohydrate, vitamins and mineral content (Ali et al., 2003) which makes it a suitable crop for ensuring food and nutrition security. However, in Uganda, the full potential of the potato crop has not been fully exploited given the lack of value-addition technologies. Value addition is limited to the consumption of fresh tubers; 92.6% as boiled potato (Tesfaye et al., 2010), and about 30–40% of potato output is non-marketable small-size tubers (rejects), a loss of US$13.7–18.3 million per annum (Kajunju et al., 2021). One MSc student assessed the possibility of increasing storability and market opportunities through the production of ready-to-fry potato strips. Another MSc student examined the feasibility of producing potato flour using low-cost technologies such as solar drying, while another MSc student examined the profitability and commercialisation potential of processing potato-based high-value food products.

FIGURE 13.1 Students working with farmers in the field

TABLE 13.1 Postgraduate students and areas of research

Name	*Research topic title*
Monica Kigambo, MSc	Effect of seed sources, seed tuber size and fungicide seed treatment on the performance of potato in Uganda. Unpublished Master's Thesis, Makerere University
Justine Nakibuule, MSc	Response of potato to manipulation of row spacing, fertilizer use and intercropping with beans. Unpublished Master's Thesis, Makerere University
Heri Bahati-Kajunju, MSc	Processing traits of major potato varieties grown in Uganda for potential production of French fries. Unpublished Master's Thesis, Makerere University. http://hdl.handle.net/10570/8360
Anthony Kwehangana, MSc	Production of Irish potato flour as a potential ingredient for food processing industry. Unpublished Master's Thesis, Makerere University
Masiko Mahafuzi, MSc	Potato cutting propagation for optimum growth and yield in Kabale, Uganda. Unpublished Master's Thesis, Makerere University
Bridget Nantambi, MSc	Consumer acceptance and willingness to pay for potato-based products: a case of urban consumers in South-western Uganda. Unpublished Master's Thesis, Makerere University
Roland Ainebyona, PhD	Potato (*Solanum tuberosum*) crop intensification innovation: a case of smallholder farmers in South-western Uganda. Unpublished PhD Thesis, Makerere University

Given these various potential interventions in quality seed production and supply, crop intensification, and value addition, there was a need for integrated approaches to help farming communities work towards achieving sustainable development goals. Roland Ainebyona, the PhD student, is analysing the efficiency and effectiveness of multi-stakeholder integrated approaches that promote the adoption of interventions and technologies developed by the six MSc students at different segments of the potato value chain.

The potato CARP approach

It was anticipated that addressing the key challenges of the potato value chain in an integrated approach would stimulate increased productivity, incomes, and food and nutrition security outcomes of the farm households who were the main target of this project. This project thus had four main interlinked components (Figure 13.1): (1) the seed potato value chain, (2) crop intensification production system, (3) postharvest handling and value addition, and (4) strengthening market linkages among the value chain actors. Component 1 was to empower farmers in seed production in order to make quality seed more available and accessible. Component 2 was to enable farmers to adopt an optimised system of potato intensification with an objective of increasing farm resource use efficiency (especially labour and land use) and increasing productivity. This would increase the demand for quality seed thereby looping into component 1. Component 3 was to empower value chain actors in the use of good postharvest handling technologies to reduce potato loss (physical and monetary), provide innovative value-addition processes that increase the value of potato, and create market linkages for the value-added products. This would motivate farmers to intensify production by optimising the use of their resources (component 2) and demand more quality seed (component 1).

The project team constituted staff and organisations with multidisciplinary backgrounds and relevant experiences. The team members were academia and researchers at Universities (Makerere University and Gulu University), Research and Extension organisations (KAZARDI and MAAIF-Local Government), a Community-Based Organisation (Excel Hort Consult), a potato business incubator and processor (Uganda Industrial Research Institute), and a private sector organisation (Agromax (U) Ltd). The team also had students (one PhD, six MSc and six BSc Interns) registered at Makerere University and 137 Diploma students at Bukalasa Agricultural College. As much as each partner had a leading role, the community-action research approaches used were participatory, demanding active involvement of the target group (ware and seed potato farmers, potato growers' associations, processors, consumers, and TVET Instructors).

The project used methodologies that enabled the potato value chain actors, especially the farmers, to actively participate in the research processes. This, in turn, enabled the project to come up with "farmer-developed and owned" alternative

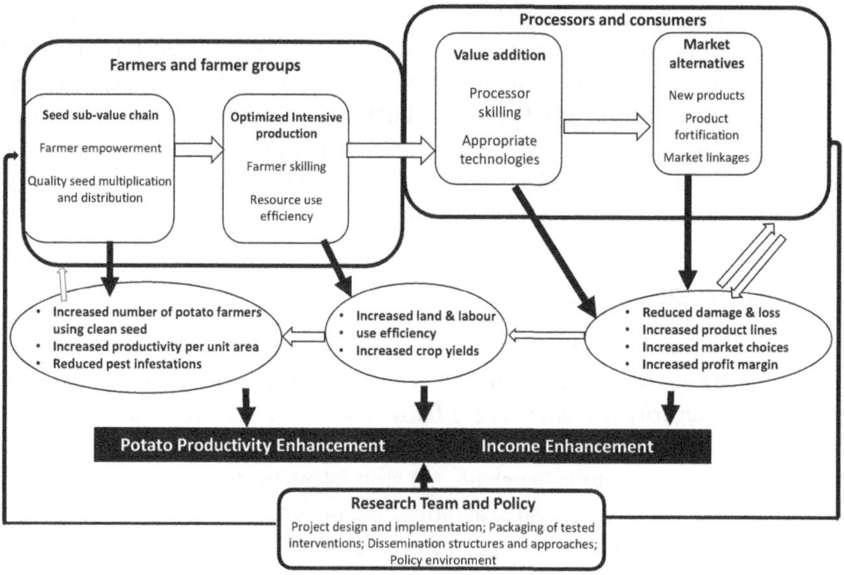

FIGURE 13.2 Conceptual framework for potato value chain to enhance farm productivity and incomes

FIGURE 13.3 In the field with farmers and with TVET students

interventions for the value chain actors to choose from, accelerating the adoption of the chosen alternative. The project designed and implemented: (1) On-farm experiments, (2) Farmer/Community Field Demonstrations (modified Farmer Field Schools), (3) Laboratory based work, (4) Technology development/incubation, and (4) Market studies.

The Farmer Field Demonstrations were set up in Bukalasa (at a TVET in Central Uganda), Mbarara, Kabale, Rubanda, and Rukiga in South-western Uganda. In agreement with the farmers, the selected demonstration sites were easy to access by the target group. In these fields best-bet crop intensification technologies and practices were demonstrated and training in good production and postharvest practices was offered. In addition, on-farm experiments were carried out based on the technologies and practices – quality seed, crop protection, fertility management, and others – that were known to have worked elsewhere, for example in Rwanda (Kathiresan, 2011) and Indonesia (Sugiarto et al., 2013). Furthermore, laboratory work was undertaken in developing and testing new value-added potato products including frozen French fries, cookies, waffles, and cakes. The focus was on methods that minimise physical losses and increase the economic and nutritional value of the products. The developed products underwent incubation targeting the youths from TVET institution and Makerere University. The potato-based products were tested for consumer acceptability and market viability.

The project first conducted a situational market analysis of the developed products in terms of their supply and demand, and the potential actors including their knowledge and perceptions on production and use/consumption of the products. Potential market demand for the products was estimated, and socio-economic and institutional factors that could influence the supply of and demand for the products were determined. Market studies were done to determine consumer willingness to buy the products. The economic viability of commercialising the products was determined. Thirty-two young entrepreneurs were trained in collective marketing, standards, market dynamics, and market positioning. The MSc and PhD students were involved in all the activities from which they generated data for their theses, journal publications, and conference papers. Four undergraduate students were also involved in supporting the graduate students mainly in data collection and community engagement.

Results

Community-based system for farmers to participate in the production, delivery and use of good quality seed potato

In a community participatory engagement, cost-effective and gender-sensitive technologies and practices that produce quality seed were identified, building on the current knowledge and practices of the farmers. Results from the farmer field demonstrations indicated that certified seed indisputably has superior performance compared to seed from informal sources (farmer-saved and local market). Large-size seed (35–55 mm) exhibited significantly high yield performance compared to small-size seed (Kigambo, 2023). Treated small-size certified seed can also be used specifically for ware potato production. Fungicide seed treatment significantly improved the number and yield of medium-size tubers besides total tuber yield. Highland/cool climate areas resulted in better yields than mid-altitudes/mid-temperate areas. Results further indicated that it is physiologically efficient and economically viable to grow potato using seed tuber cuttings. For maximum survival of the cut seed in the soil, the cuts must be treated with mancozeb (a non-systemic fungicide) as a protectant against seed decay.

System of potato intensification (SPI) that optimises farmers' resources

From the on-farm community participatory experiments, results showed that intercropping potato and beans contributed to tuber quality, yield, and efficient land use (Nakibuule et al., 2022). The results also dismissed the perception that only monocrops give high yields, and tailoring thinking to diversification – a viable mitigation of climate change effects. Intercropping potato and beans suited commercial potato production with the least wastage (reduced quantity of tiny non-marketable tubers). The study highlighted that most on-farm soils were depleted of essential nutrients.

Postharvest technologies to minimise potato postharvest losses and increase marketability

Determination of physico-chemical characteristics of the major potato varieties grown in Uganda revealed that the majority of the varieties have high dry matter content (>20%), with tubers that are medium in size (50–60 mm), round in shape with medium eye depth (Kajunju et al., 2021). The varieties have fairly good processing quality traits with Kachpot1, Kinigi, and Rwashaki being the most suitable for processing French fries due to high starch content (>20%), low reducing sugar concentration (less than 0.50g/100g FWless than 0.50 g100 g-1 FW), and big and round tubers with medium eye depth. The other potato varieties such as Cruza, Kimuli, Rwangume, Mbumbamagara, Rutuku, and Victoria are appropriate for domestic preparation of potato flour, mashed potato and salads (Kajunju et al., 2022). Based on this information, the project developed and tested a number of potato-based value-added products, namely, potato flour and ready-to-eat snacks including cookies, waffles, chapatti, bread, and cakes (Kwehangana, 2022). There is empirical evidence that the manufacture of these products is feasible and economically viable. They have low sugar and gluten levels (gluten even more so), while others are gluten-free. They are rich in dietary fibre, potassium, calcium, iron, zinc, and vitamins A and C.

Market potential of the potato value added products and promotion of market linkages for the products

The market studies conducted helped the project to identify market niches and strategies that potential entrepreneurs could adopt to penetrate and sustain the market. Consumers were willing to pay about 1,600 Uganda shillings (US$0.46) per 45g pack of potato-based cookies. This is an indicator of the likely market entry success given that the producer would break even by selling the cookies at 1,000 Uganda shillings (US$0.29). The price of close substitutes (other cookies and biscuits already in the market) and the distance to market, that is, where consumers buy the cookies, were the key factors that would influence the price consumers would be willing to pay for the products. The products have acceptable attributes (shape, colour, taste, etc.), hence a high potential of entering the market. Six products were tested and validated by the authorised national body. In order to ascertain that the developed products are commercialised, the project identified and trained those entrepreneurs with the potential to undertake commercial production. Two business companies owned by the graduate students on the project were created. This created linkages between the University, researchers, and the private sector and raised the employability of the students.

(a)

(b)

FIGURE 13.4 Adding value: (a) potato flour waffles; (b) toasted sandwiches

Impact

The project envisaged benefiting potato value chain actors, primarily smallholder farmers, processors, and consumers in Uganda. For wide reach, the project adopted a group approach with the anticipation that the farmers who participated in demonstrations and experiments were able to share with others in the group. The participation of farmers has resulted in the adoption of an optimised potato intensification system, increasing farm resource use efficiency, and productivity for both potato

(a)

(b)

FIGURE 13.5 Students baking products and exhibiting them: (a) cookies; (b) potato snacks

and other crops. In addition, the use of innovative value-addition processes has increased the value of potato. A number of potato-based processed products are now available to consumers. The overall impact is enhanced sources of farmer and community livelihoods with increased capacity of potato value chain actors to drive the chain. Food and nutrition security has also been positively affected, especially for those who are wheat intolerant.

The project benefited the TVET Institution by skilling six teaching staff and 61 students. The project brought on board the academic staff of the TVET Institution, who were initially non-research oriented. It enhanced their research skills and strengthened the relationship between the TVET and the University. The inclusion of the TVET was to facilitate the learning of agricultural skills and to support closer links with the University. The students at Diploma and Certificate levels were trained in different aspects of the potato value chain with a focus on good agricultural practices and entrepreneurship. The involvement of the students in the project activities, together with the training, enhanced the spread of agricultural knowledge and technologies.

Other beneficiaries are University students attached to the project. The 11 university students (one PhD, six MSc and four BSc interns), in addition to conducting research within the project framework, undertook facilitation of the farmer learning processes. Since their graduation, the students have been applying the acquired research and theoretical skills to serve the agricultural sector at different nodes of the value chains. Muhingi Products Ltd. is one of the successful stories from these students. Muhingi Products is an SME created by the students on the project. The company processes potato-based products including gluten-free cookies, employs 11 people, and collaborates with over 100 farmers to source raw materials.

One of the most successful Farmer Field Demonstrations was set up in the Diocese of Kigezi because of the number of people that could access it, and also because of the commitment of farmers and the ease of access by a wide community. This arrangement impacted three farmer groups (each of 33 farmers) who directly participated in running the demonstrations. The approach was so cost-effective that other non-group farmers totalling 1,729 farmers were reached through different project activities. In addition, the training of potential entrepreneurs has led to the establishment of two companies that have adopted the processing and commercialisation of potato-based products.

At an institutional level, the visibility of the project researchers and Makerere University at large has increased, springing from the strengthened relationship with the farming community, the private sector, and the local government. The government of Uganda is increasingly interested in the project, for instance, the government, through the Makerere University Research and Innovation Fund, has provided supplementary funds to expand the crop coverage of the platform; and the Multi-Sectoral Food and Nutrition Security Project of the Ministry of Agriculture, Animal Industry and Fisheries has expressed interest in a partnership.

Conclusion and challenges

The potato crop value chain in Uganda was enhanced for improved livelihoods of the farming communities and other value chain actors through multidisciplinary participatory approaches. An improved seed value chain was strengthened that enables the farmers to access quality seed and empowers them in seed selection and handling impacts on crop productivity and profitability. Similarly, the adoption of optimal crop intensification practices impacted positively on the quality and yield of crop harvests, hence increasing the use efficiency of scarce farm resources. One of the viable market linkages for farmers' potato output was initiated by exploring value addition and processing. This CARP showed that the production of high-value processed products is feasible and economically viable. The processed products have attributes that are acceptable to the market, and consumers are willing to pay a price that enables the producers to break even. Commercialisation of the products also creates backward linkages in the value chain that will impact positively on farmers' productivity and incomes.

The use of an innovation platform was difficult to implement initially but showed itself to be very worthwhile despite all the time involved. It has contributed to enhancing the network of the Principal Investigator, all the supervisors and the students, and helped to strengthen the relationship between the University, a TVET, and the private sector, mainly the farming community. The most challenging was maintaining cohesion among different stakeholders of the innovation platform, particularly because they had different goals and different expectations from the project. This was worsened by the outbreak of the COVID-19 pandemic that cut off the partners, the target group and the students from one another for over four seasons. Nonetheless, the Principal Investigator, who has researched potato for about two decades, achieved great satisfaction having generated working solutions for the Uganda potato value chain.

Interested readers are referred to the RUFORUM Impact Platform www.ruforum.org/impact/ to see a film on the Potato CARP.

References

Ali, Asghar, Abdur Rab, and Syed Asghar Hussain. 2003. 'Yield and Nutrients Profile of Potato Tubers at Various Stages of Development'. *Asian Journal of Plant Sciences*, 2(2): 247–250. https://doi.org/10.3923/ajps.2003.247.250.

Arinaitwe, Abel Byarugaba, John Kyooma, Ambrose Aheisibwe Rwaheru, Deo Tibanyedera, and Barekye Alex. 2017. 'Bridging the Gap in Quality and Quantity of Seed Potatoes through Farmer Managed Screen Houses in Uganda'. *African Journal of Plant Science*, 11(2): 30–37. https://doi.org/10.5897/ajps2016.1485.

Kajunju, N. Heri Bahati, Abel Atukwase, Gaston-Ampe Tumuhimbise, and Johnny Mugisha. 2022. 'Characterisation of Potato Varieties Commonly Grown in Uganda for Food Processing Suitability'. *African Crop Science Journal*, 30(1): 51–68. https://doi.org/10.4314/acsj.v30i1.4.

Kajunju, N. Heri Bahati, Abel Atukwase, Gaston-Ampe Tumuhimbise, and Johnny Mugisha. 2021. 'Potato Processing in Uganda: A Technical Review'. *Makerere University Journal of Agricultural and Environmental Sciences*, 10(1): 60–81.

Kathiresan, A. 2011. 'Strategies for Sustainable Crop Intensification in Rwanda: Shifting Focus from Producing Enough to Producing Surplus'. Ministry of Agriculture and Animal Resources, Kigali, Rwanda.

Kigambo, Monica. 2023. 'Effect of Seed Sources, Seed Tuber Size and Fungicide Seed Treatment on the Performance of Potato in Uganda'. Unpublished master's thesis. Makerere University, Kampala.

Kwehangana, Anthony. 2022. 'Production of Irish Potato Flour as a Potential Ingredient for Food Processing Industry'. Unpublished master's thesis. Makerere University, Kampala.

Lutaladio, NeBambi, Nicolaus Cromme, B Adam Prakash, and Fernando Ezeta. 2010. 'Strengthening Potato Value Chains: Technical and Policy Options for Developing Countries'. FAO and CFC, Rome, 2010.

Mbowa, Swaibu and Francis Mwesigye. 2016. 'Investment Opportunities and Challenges in the Irish Potato Value Chain in Uganda'. Research Report No. 14, Economic Policy Research Centre. https://eprcug.org/publication/investment-opportunities-and-challenges-in-the-irish-potato-value-chain-in-uganda/.

Mugisha, Johnny, Robert Mwadime, Christopher Sebatta, Robert Gensi, and Bernard Obaa. 2017. 'Factors Enhancing Household Nutrition Outcomes in Potato Value Chain in South-Western Uganda'. *Journal of Sustainable Development*, 10(3): 215–230. https://doi.org/10.5539/jsd.v10n3p215.

Muthoni, Jane, Hussein Shimelis, and Rob Melis. 2013. 'Potato Production in Kenya: Farming Systems and Production Constraints'. *Journal of Agricultural Science*, 5(5): 182–197. https://doi.org/10.5539/jas.v5n5p182.

Nakibuule, Justine, Jimmy Obala, Monica Kigambo, N. Heri Bahati Kajunju, and Johnny Mugisha. 2022. 'Effect of Potato-bean Intercrop Arrangement, Plant Spacing and Fertiliser Usage on Plant Growth and Tuber Yield in Different Environments'. *Makerere University Journal of Agricultural and Environmental Sciences*, 11(2): 107–126.

Sugiarto, S., Sudiarso, R., and Soemarno, S. 2013. 'Local Potential Intensification System (SIPLO) the Sustainable Management of Soil Organic Potatoes'. *International Journal of Engineering and Science*, 2(9): 51–57.

Tesfaye, A, Lemaga, B., Mwakasendo, J. A., Nzohabonayoz, Z., Mutware, J., Wanda, K. Y., Kinyae, P. M., Ortiz, O., and Crissman C. 2010. 'Markets for Fresh and Frozen Potato Chips in the ASARECA Region and the Potential for Regional Trade: Ethiopia, Tanzania, Rwanda, Kenya, Burundi and Uganda'. International Potato Center, Lima.

14

ENHANCING PIG PRODUCTION AND MARKETING FOR SMALLHOLDER FARMERS IN NORTHERN UGANDA

Elly Kurobuza Ndyomugyenyi, Tony Aliro and Walter Odongo

Introduction

Pig rearing has the potential for improving nutrition and household livelihoods in sub-Saharan Africa (SSA). With a population of 3.6 million, pig production provides livelihoods for over 1.1 million households in Uganda (FAO, 2020). The pig sub-sector also contributes about 12% of the total livestock GDP and over 3% of the total agricultural GDP in Uganda (UBOS, 2019; FAO, 2020).

The increasing interest in pig production is for its ability to provide protein in the face of climate change. Pigs have short gestation periods, rapid multiplication rates and the ability to feed on many crop and animal products as well as waste (Ndyomugyenyi and Kyasimire, 2015). Despite this, the total pig population in Uganda has not significantly increased: from 3.6 million in 2012 to only 4.2 million in 2019 (UBOS, 2015; UBOS, 2019). The 17% increase in pig population lags behind the increase in demand for pork, estimated at 3.3 kg per capita. Northern Uganda trails the rest of Uganda in having the lowest, with only 13% of the total pig population in the country (UBOS, 2019) owned by about 10% of the households (Tatwangire, 2014).

The low pig population is attributed to the high cost of feeds, poor breeds, poor pig health management, lack of suitable housing for densely populated farming areas, limited processing, and lack of reliable markets for pigs and pig products. The high cost of feed in livestock production necessitates the assessment of less costly alternatives. This project assessed the use of less costly alternative feedstuffs, such as sweet potato vines, cassava tubers and rice bran, in the diets of the growing pigs. The challenge of poor breeds was addressed by using artificial insemination (AI). The use of AI as an alternative to natural mating has the advantages

DOI: 10.4324/9781003387497-17

of improving growth rates, pig genotypes, reducing disease transmission and the cost of maintaining boars (Ndyomugyenyi and Kyasimire, 2015). However, using commercial semen extenders in AI is costly and unaffordable for smallholders (Muhanguzi et al., 2012). The CARP addressed this challenge by using locally made semen extenders as an alternative to commercial extenders. The challenge of foul smell emissions from pig houses has been addressed by more complex methods such as decomposition and the use of digesters (Chastain, 2003). These methods are usually expensive and unaffordable for smallholder farmers. The use of Indigenous Micro-organisms (IMO) have been suggested as a less costly alternative (Ndyomugyenyi and Kyasimire, 2015) but its effectiveness in local housing settings has not been fully exploited. There is also the challenge affecting the marketing of pigs in Uganda. There is a lack of understanding of key customers, their interests in terms of quantity, quality and trends in demand (Mulindwa, 2016). This arises due to inadequate access by smallholders to pig markets. Enhancing profitable market participation along the value chains strengthens competitiveness and ensures increased income status of smallholders (Ouma et al., 2016). This project identified and classified primary markets and customer segments for pigs and pig products.

This chapter documents the process and outcomes of a Community Action Research Project (CARP) aimed at enhancing pig production and marketing among smallholder farmers in northern Uganda. The project promoted pig production technologies and marketing along the pig value chain. Specifically, the project targeted local feed formulations to reduce feeding costs, use of indigenous micro-organisms (IMO) as a pig health management approach, use of artificial insemination (AI) to improve pig genotypes, the promotion of value addition, and the improvement of farmer collaboration and marketing options for pigs and pork products.

FIGURE 14.1 Working with farmers to establish the baseline

Approach

The CARP was conducted following a participatory approach to research and development. A multi-stakeholder platform was established consisting of Gulu University working closely with a Technical, Vocational Education and Training (TVET) institution, smallholder pig farmers, the private sector, the local government extension department, and others, including a local community organization. The platform was used to stimulate innovations, research, experience sharing and joint learning.

A coherent platform was achieved with support from Local Government and the willingness of all stakeholders to participate in project implementation. Following the inaugural meeting, a baseline survey was conducted to establish the status of the pig value chain. The design and description of the activities, outputs, outcomes and impact are summarized in Figure 14.2. The participatory approach during planning, inception and project implementation enabled buy-in and helped to address the concerns related to the expectations of different stakeholder groups. Five MSc. students and one PhD student were selected and fully sponsored and an additional four MSc. students were partially sponsored through CARP activities. The research methods used for the student research projects are shown in their theses and papers. Mobilization of farmers was done by the Community Knowledge Workers of the Local Government. The farmers were directly engaged in all students' research activities. The TVET institution provided preliminary experimental sites and was engaged in training its students and future community agricultural extension agents using the innovations.

Gulu University was supportive of the CARP approach which reinforces the Gulu University vision: "For Community Transformation". Additionally, the CARP easily fitted within the University's student-centred outreach model that champions University–community engagement. The support enabled timely implementation and reinforced completion of the project activities, as well as the students' research and graduation.

Results

The CARP platform results are categorized into two thematic areas based on the focus of the project, that is, production innovation – including pig health management – and marketing and processing. Most of these findings are action oriented and have already been put into practice by farmers. Scientific articles were also developed in international peer-reviewed journals for the academic community. The key findings are briefly summarized in the sections below.

Pig production technology innovations

Use of indigenous micro-organism technology improves pig performance and health

The first research investigated the effect of deep litter floor treated with indigenous micro-organisms (IMO) on pig production and the nutrient content of pork. The

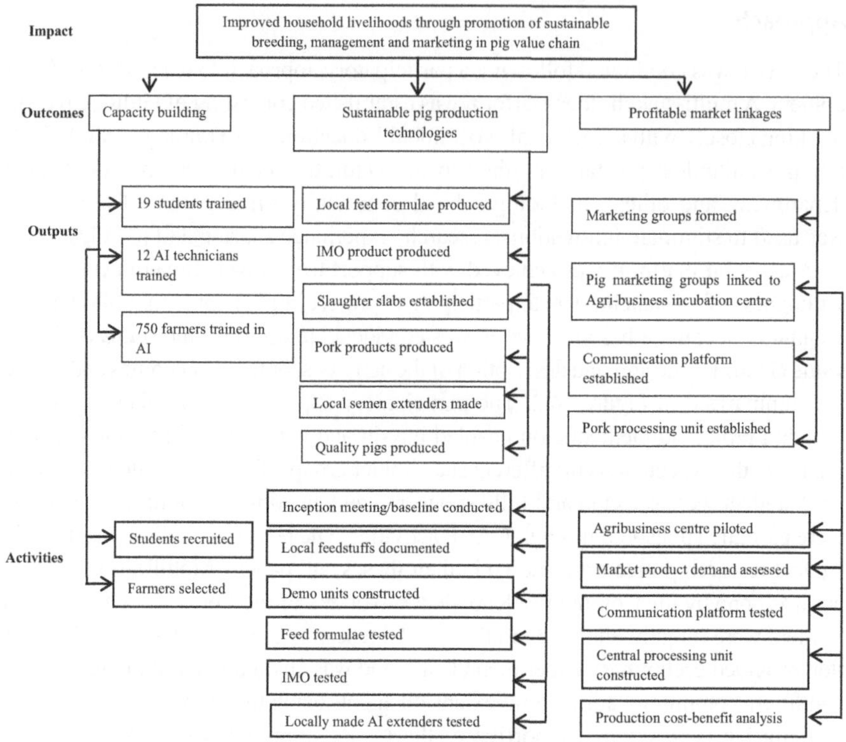

FIGURE 14.2 Pig CARP theory of change as implemented

results showed that pigs raised on IMO-treated deep litter floor had better weight gain (60%) and better-quality pork with up to 53% unsaturated fatty acids compared to those raised on untreated deep litter floor (weight gain, 40%; saturated fatty acids, 47%). Two articles were published from this research (Kidega et al., 2021; Kidega et al., 2020). The second action research investigated the effect of IMO treatment of the deep litter floor on the reduction of foul smell. The results showed that odor from treated deep litter houses could not be smelt within a radius of 10 meters, suggesting that IMO is effective in reducing foul smell to environmentally friendly levels. The use of IMO also reduces the time spent on other management practices like cleaning, disinfection, and treatment of external and internal parasites. The effectiveness of IMO is critical in pig production. It also improves pig health which is important. Some 78% of smallholder farmers had sought veterinary services for the management of several pig health conditions and vaccination. From this research project, two articles were published (Okello et al., 2022; Mamawi et al., 2020).

Collaboration with the TVET institution and engagement of the students with farmers led to a series of training on how to establish the appropriate houses, how to use local materials, and on how to use the IMO. There was rapid uptake and farmers were working together to build the necessary shelters before the research on productivity was completed. By the end of the project, some 80% of farmers interviewed were using IMO.

Use of local feed formulations reduces pig feeding costs

Research was conducted with farmers to investigate the diversity, effectiveness and the economics of local feed resources for pig production. The research involved assessing a wide range of combinations to determine the best formulae. The results indicated that common easily available ingredients on the farms included cassava, sweet potato vines, rice bran, pig weed, wandering Jew, and maize bran. Of the formulae produced, most were effective, and the best was a feed that included a mixture of sweet potato vines, cassava tubers and rice bran. Formulated commercial feed is more costly (963 Uganda shillings per kg) than formulated local feed (213 Uganda shillings per kg). Over 76% of smallholder farmers were using local feed resources in pig production by the end of the project (Okello et al., 2021). An analysis of the effects of substituting commercial feed with fermented vegetable peelings and crop waste indicated the potential for increasing returns from pig production (Kasima et al., 2023).

Use of artificial insemination (AI) in pig production

The research assessed farmers' awareness and perception of AI technology and the effectiveness of using coconut water as a semen extender. Results showed that over 70% of pig farmers were aware of AI, although none had used it before. The

FIGURE 14.3 Farmers working together to prepare IMO-appropriate pig house

perceived usefulness and attitude toward AI were critical drivers for the intention to use AI technology. It was established that using coconut water can extend semen for 24 hours at room temperature, and for 96 hours at 17–19°C. Insemination experiments by AI technicians yielded positive results with an average litter size of 10 piglets per sow, and a conception rate of 85.4%. This suggests that coconut water-based semen extender can be sustainably adopted and scaled up. Further research is urgently needed to standardize this semen extender before it is adopted and scaled up for pig breeding. An article about this research was published in University World News.

Innovations in marketing and effective information sharing

Research investigated the determinants of innovation behavior among smallholder pig producers and consumer preference for pork products. Results showed that the farmer innovation process is not a continuous uninterrupted process but rather a random one where improved access to market information enhances farmer innovation behavior. Pork consumers had a high preference for ham and ribs, while freshness, taste and aroma were the highly ranked attributes of pork during purchase. Understanding consumer quality demands enhances farmer innovation by enabling a better understanding of the changing market demands so as to devise coping strategies and ensure better decision-making power. Two articles were published from this research (Mugonya et al., 2021; Mugonya et al., 2020).

CARP outcomes, impact and lessons learned

Capacity building and employability of students

Commitment and effective engagement of students in the project enabled seven MSc. out of the nine (78%) students to graduate on time with eleven articles published. The engagement of students in the CARP increased their employability and commitment to community transformation. Most are either employed in non-governmental organizations (NGOs), self-employed, or both. The project developed the capacity of 97 students which included one PhD and eight MSc. (Table 14.1) with support from ten BSc. and 78 TVET institution students. It trained 300 farmers in IMO and feed production and ten farmers in pork processing and value addition. It also trained 12 AI technicians. Seven supervisors and researchers were engaged in the project.

Training improved the employability of the graduate students and their ability to further their studies. Junior Kasima Senyonga is working as an instructor in the Department of Animal Production at the International College of Agricultural Sciences and a consultant at Apex Research Solutions, Uganda. In his free time, he makes leather shoes for extra income. He bought a piece of land and intends to set up a pig farm. John Mugonya is Program Manager at Agripreneurship Alliance,

TABLE 14.1 Postgraduate students and research topics

Degree	*Student name*	*Thesis title*
MSc.	Junior Kasima Senyonga	Effect of floor type and local feed formula on pig production in Gulu and Omoro districts, Northern Uganda
MSc.	Lucy Aciro	Determinants of consumer preference for pork products in Northern Uganda
MSc.	Sarudzai Muzhange	Economics of local feed resources for pig production in Northern Uganda
MSc.	Prima Kyohairwe	Perceptions of artificial insemination in pig production in Northern Uganda
MSc.	Kenneth Kidega	Effect of deep litter floor housing system on performance of pigs and pork quality
MSc.	Godfrey Mamawi Ambayo	Effectiveness of indigenous microorganisms in controlling odour in pig houses
MSc.	John Mugonya	Determinants of innovation behaviour among smallholder producers of pigs in Northern Uganda
PhD	Daniel Micheal Okello	Production, market orientation, innovation and performance among smallholder pig farmers: The case of Northern Uganda
MSc.	Julian Lawach	Hydroponic fodder as feed for weaned piglets in Gulu and Omoro districts

Uganda, responsible for staff management, budgeting, financial management, research, and fundraising. Kenneth Kidega is self-employed. He breeds sows and sells off weaned piglets. Lucy Aciro is employed as an Assistant Lecturer and Quality Assurance Officer at the University of Sacred Heart, Gulu. She is now admitted on a PhD program in Agrisystems at the University of Cattolica in Italy. Sarudzai Muzhange is self-employed, running a butchery where pork is one of the products. Geoffrey Mamawi Ambayo is employed with District Local Government as District Production Officer. He effectively manages Parish Development Model, engages in community mobilization, and manages multi-sectoral community projects. He is now admitted on a PhD program in Agricultural and Applied Biosciences at Gulu University. Daniel Micheal Okello had published three articles from his studies before submitting his thesis.

Improved uptake of innovations developed by the CARP in communities

The CARP engaged nearly 1,000 smallholder farmers. Farmers under the project have been trained in better husbandry practices and are now organized into a pig farmers' association, something that never existed in the area. The uptake of CARP innovations has enabled farmers to adopt better husbandry practices. At the beginning of the project, none of the farmers was using AI or IMO for pig production.

FIGURE 14.4 Celebrating at training session with the community

As a result of the project, more farmers have improved pig genotypes using AI and housing conditions using IMO. In a survey conducted in October 2022, 86 out of 96 farmers interviewed had constructed new IMO structures, while 45 farmers were using AI for pig breeding. Overall, there is more willingness to try out new technologies among pig farmers engaged in the CARP as a result of working together on the research, the demonstrations and dissemination of results during project implementation. They now see the University as a partner in their efforts to improve agricultural production.

CARP established stronger links between the University and communities

The CARP has enabled Gulu University to create strong linkages with TVETs, CBOs, private sector and other research institutions such as Makerere University and the International Livestock Research Institute (ILRI). Through Gulu University's Community Student Attachment program, students have been attached to CARP collaborators for internships and experiential learning, hence broadening attachment opportunities and cross-learning. The linkages have aided students to identify the gaps/problems in the communities for their research projects on a wide range of farmer issues, hence promoting applied research. The linkages have opened channels for farmers and TVETs to freely access the University, enabling joint development of demand-driven, short training and competence-based curricula for TVET students, University students and pig farmers. Ultimately, the University has become more relevant and visible to the community. The cooperation established with research institutions such as ILRI and Makerere University could lead to potential new engagements, particularly in areas of research and student supervision. The engagements also established relationships and a platform for dialogue with policymakers, as the local governments and other government agencies were involved throughout the research phase and the dissemination activities of the CARP. Finally, the farmers became organized into groups which strengthened their bargaining power. Market linkages for pig farmers were created and a processing facility was established at the University that is accessible to the pig value chain actors. These initiatives have also strengthened the University–community engagement network of Gulu University.

CARP improved welfare of pigs and livelihoods of communities

Farmers, particularly those who housed CARP demonstration units, are role models. Farmers from other regions come and learn better pig production practices such as the use of IMO deep litter floor and local feed formulae. Consequently, some farmers have benefited from being role models by attaining high political positions. One such example is Mr Charles Okello who is currently the Local Council Chairperson of Paicho sub-county. In the past farmers did not often house pigs, thus not producing manure to be available for crops. Due to the presence of pig manure from IMO deep litter floor, there has been diversification of enterprises among farmers, especially for horticultural and sunflower production. There has been improved marketing where farmers sell value-added pig products. Instead of selling live pigs, farmers slaughter and sell either fresh or deep-fried pork. Through earnings from pigs and pig products, some farmers have constructed residential or commercial houses, taken children to schools, and improved housing conditions for pigs. Currently, all farmers provide shelter to their pigs compared to the 59.6% who sheltered the pigs before the CARP intervention.

CARP incubation centre improved learning, research, and business incubations

Farmers who were trained to process and add value to pork are currently using the CARP incubation facilities to produce pork products for sale. These farmers are earning 100,000 Uganda shillings (US$27) per day. They are expected to develop their independent businesses to allow other farmers to benefit as well. In fact, they are in the process of certifying the products. The incubation centre is also an

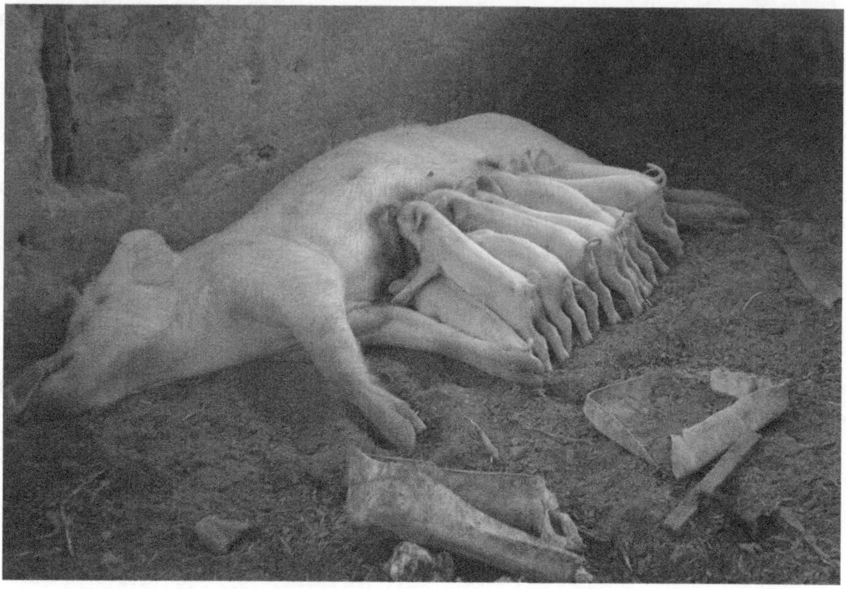

FIGURE 14.5 Contented and smell-free pigs

avenue for student practical skills training and research. One student is working on the inclusion of vitamins and antioxidants to enrich the quality of pork sausages. Thus, the CARP has contributed to the transformation of the University through research, student supervision, and publications. Through the engagement of farmers and students in the research and production of value-added pork products, the profile of the University in society has improved.

CARP has connected various stakeholders to strengthen the sustainability of the project and stimulated the spread of some innovations

The CARP has brought uncoordinated smallholder farmers together to explore the wider market for their pigs, first among themselves and, secondly, selling as a group. Subsequently, farmers have been organized into an association, which strengthens the sustainability of the CARP initiatives. The CARP has enabled the writing of collaborative research proposals with other universities as a result of connections obtained from attending RUFORUM Annual and Biennial Conferences. The engagement of AI technicians from the District Production Department has enabled the continuity of pig AI, particularly in the Parish Development Model program. The pig processing and value-addition incubation centre has enabled farmers to incubate their value-added businesses, hence contributing to their livelihoods.

The exchange of ideas among students has resulted in a student from Cameroon, who was not part of the project, taking the ideas back home after graduating where he has set up a pig production and marketing business, adopting these technologies and approaches.

Lessons learned from CARP implementation

CARP has facilitated pig value chain shortening for the benefit of farmers

As a result of the CARP intervention, some of the smallholder pig producers who usually sell live pigs as individuals or groups are currently processing the pigs into fresh pork or deep-fried pork. In this way, they are earning more money than selling live pigs. One successful early adopter (Mr Geoffrey Odoch) has benefited from the value chain shortening. Instead of selling live pigs, he slaughters them and sells fresh or deep-fried pork. Mr Odoch also sells fresh pork to Gulu City pork sellers and farmers at CARP pork processing incubation centre. He gets a profit of 600,000 Uganda shillings (US$174) per day from the three pigs he slaughters. Due to the high demand of pork, Mr Odoch has grown seven acres of maize and five acres of sunflower to reduce the cost of feeding. He has innovatively engaged out-growers who supply him with more sunflower seeds. He links with oil extraction businesses where he sells the oil at an agreed price but retains the cake for feeding pigs.

Unlike other farmers in rural settings where the extent to which value addition to their products is limited to only primary processing, Mr Odoch has created a link between farmers at the CARP pork processing incubation centre who buy fresh pork from him for further value addition (secondary processing). These farmers have reached a stage of certifying their value-added pork products. Successful early adopters such as Mr Geoffrey Odoch provide good success stories for motivating others. For instance, out-of-school youth having seen this development are working with Mr Odoch to extend it to the community.

CARP has facilitated mindset change toward new innovations

For innovations to be successfully adopted, farmers need to be convinced that the innovations indeed work. Most farmers considered exotic practices better than those that are indigenous. In particular, pig production has not contributed much to the livelihoods of most farmers because of the high cost of feeds; AI has not been used much because it was costly. However, CARP experiments showed that locally available feedstuffs produce low-cost feeds of the same quality as commercial feeds. In this way, the attitude of the farmers toward the use of locally formulated feeds changed. Likewise, lack of access to AI services has resulted in disease transmission as one boar can be used by the whole village to breed sows. Farmer appreciation of AI technology, because it is more affordable with the coconut extender, has enabled many farmers to take up the technology. Three AI businesses integrated as part of veterinary services have been established with students and are expected to serve a total of 522,000 persons in the districts of Gulu (325,600) and Omoro (196,400). In respect to IMO technology, farmers thought that using concrete floors in pig housing was superior. However, the mindset of farmers changed after the introduction of IMO technology by the CARP. The results showed that IMO deep litter floor using local materials was affordable and not only reduced the foul smell from pig houses but also improved pig production and pork quality.

Conclusion

The CARP approach was a big departure from a classical research approach as it required the research team to generate research ideas based on actual societal challenges as identified jointly with the pig value chain actors. Secondly, it involved setting up and operationalization of a coordinated trans-disciplinary research approach involving many strands such as technology and innovation development, innovation testing, market research, and community outreach. This approach strongly positions the University to offer solutions to local challenges through scientific research. It also builds the capacity of researchers and students as community change agents, while integrating entrepreneurship orientation in learning and research.

Despite the advantages associated with this approach, it requires significant additional effort by the Principal Investigator to facilitate and manage the diverse activities and stakeholders. But the results of engaging in the platform have benefited

FIGURE 14.6 Farmers feeding surplus and waste products to pigs

all those involved. This approach should be scaled out and deepened within the University and beyond. The sustainability of this platform also requires substantial further investment, especially to strengthen the incubation hub at the University and with the TVETs. Ongoing engagement with all stakeholders would enable the identification of emerging challenges and enhance consolidation of the impact of the multi-stakeholder platform and its influence in the value chain business environment.

Interested readers are referred to the Pig CARP film on the RUFORUM Impact Platform https://ruforum.org/impact/project/gulu-pig-carp-2021/.

Bibliography

Chastain, J. P. 2003. 'Air quality and odour control from swine production facilities'. https://www.clemson.edu/extension/camm/manuals/swine/sch9_03.pdf.

FAO (Food and Agriculture Organization). 2020. 'FAO statistical data base'. http://www.fao.org/faostat/en/#data.

Kasima, Junior Senyonga, Basil Mugonola, and Elly Ndyomugyenyi 2023 'Effect of substituting a commercial grower diet with fermented sweet potato vine-cassava tuber-based diet on the performance of growing pigs'. *African Journal of Rural Development*, 6 (2): 44–52. https://afjrd.org/index.php/jos/article/view/301.

Kidega, K., Ndyomugyenyi, E. K., and I. Okello-Uma, I. 2020. 'Effect of indigenous micro-organism treatment of deep litter floor on nutrient content of pork'. *African Crop Science Journal*, 28(s1): 247–254. DOI: https://dx.doi.org/10.4314/acsj.v28i1.19S.

Kidega, K., Aliro, T. Mugonola, B., Ndyomugyenyi, E. K., and Okello-Uma, I. 2021. 'Effect of different floor types on growth performance of pigs and carcass back-fat thickness'. *London Journal of Research in Science: Natural and Formal*, 21(2): 47–54.

Mamawi, A. G., Aliro, T., Kidega, K., and Ndyomugyenyi, E. K. 2020. 'Effectiveness of indigenous micro-organism treatment of deep litter floor in pig house foul odour control'. *IAR Journal of Agriculture Research and Life Sciences*, 1(4): 113–120.

Mugonya, J., Kalule, S. W., and Ndyomugyenyi, E. K. 2021. 'Effect of market information quality, sharing and utilisation on the innovation behaviour of smallholder pig producers'. *Cogent Food and Agriculture*, 7(1). DOI: https://doi.org/10.1080/23311932.2021.1948726.

Mugonya, J., Kalule, S. W., and Ndyomugyenyi, E. K. 2020. 'Utilisation of labour among pig farmers in northern Uganda'. *African Crop Science Journal*, 28(s1): 237–246. DOI: https://dx.doi.org/10.4314/acsj.v28i1.18S.

Muhanguzi, D., Lutwama, V., and Mwiine, F. N. 2012. 'Factors that influence pig production in Central Uganda: Case study of Nangabo Sub-County, Wakiso district'. *Veterinary World*, 5(6): 346–351. DOI: 10.5455/vetworld.2012.346–351.

Mulindwa, C., 2016. 'Training guide: Pig and pig meat marketing in Uganda'. http://www.rtb.cgiar.org/wp-content/uploads/2015/06/Guide-on-Pig-Marketing-Final.pdf.

Ndyomugyenyi, E. K., and Kyasimire, J. 2015. 'Pig production in Kichwamba Sub-county, Rubirizi district, Uganda'. *Livestock Research for Rural Development*, 27(10). http://www.lrrd.org/lrrd27/10/kuro27199.htm.

Okello, D. M., Odongo, W., Aliro, T., Ndyomugyenyi, E. K, and Owiny, D. O. 2022. 'Alone or a combination: Ascertaining factors associated with choice of pig health management strategies amongst smallholder farmers in northern Uganda'. *Preventive Veterinary Medicine*, 199. DOI: https://doi.org/10.1016/j.prevetmed.2021.105562.

Okello, D. M., Odongo, W., Aliro, T., and Ndyomugyenyi, E. K. 2021. 'An assessment of pig feed diversity amongst smallholder pig farmers in Northern Uganda'. *Cogent Food and Agriculture*, 7(1). DOI: https://doi.org/10.1080/23311932.2021.1937867.

Ouma, E., Ochieng, J., Dione, M., and Pezo, D. 2016. 'Governance structures and constraints along the Ugandan smallholder pig value chains'. 5th International Conference of the African Association of Agricultural Economists, September 2016, Addis Ababa, Ethiopia. http://ageconsearch.umn.edu/handle/246976.

Ouma, E., ione, M., Lule, P., Roesel K., and Pezo, D. 2013. 'Characterization of smallholder pig production systems in Uganda: Constraints and opportunities for engaging with market systems'. *Livestock Research for Rural Development*, 26(3): 56.

Tatwangire, A. 2014. 'Uganda smallholder pig value chain development: Situation analysis and trends'. International Livestock Research Institute. Nairobi, Kenya. https://hdl.handle.net/10568/34090.

Thu Hong, T. T. and Thanh Ca, L. 2013. 'The protein content of cassava residue, Soybean waste and rice bran is increased through fermentation with Aspergillus oryzae'. *Livestock Research for Rural Development*, 25(7). http://www.lrrd.org/lrrd25/7/hong25132.htm.

UBOS (Uganda Bureau of Statistics). 2015. 'Statistical abstract. Kampala, Uganda'. http://library.health.go.ug/sites/default/files/resources/UBOS%20Statistical%20Abstract%202015.pdf.

UBOS (Uganda Bureau of Statistics). 2019. 'Statistical abstract. Kampala, Uganda'. http://library.health.go.ug/monitoring-and-evaluation/statistics/ubos-statistical-abstract-2019.

University World News. 2022. 'Coconut water aids artificial insemination of pigs'. https://www.scidev.net/sub-saharan-africa/news/coconut-water-aids-artificial-insemination-of-pigs/.

15

BUILDING COMPETITIVENESS FOR COMMUNAL FARMERS THROUGH DEVELOPING THE WOOL VALUE CHAIN IN THE FREE STATE PROVINCE OF SOUTH AFRICA

Jan W. Swanepoel

Introduction

The global wool industry is currently in a privileged position and has ample opportunity for growth. Wool prices are stabilising while economists and wool buyers believe this trend will continue due to international market demand exceeding supply. The creation of niche products from wool adds to the existing value chain, creating more jobs and an opportunity for enlarging the export market within the South African economy. Even though the wool industry has provided ample growth opportunities, small wool growers experience difficulty in obtaining maximum income from wool.

In the past, communal livestock farmers, predominantly wool farmers, were not integrated into the mainstream agricultural economy of South Africa. This caused them difficulty accessing proper markets, resulting in lower prices for their wool. The low quality and lack of management skills further increased their lack of participation in the mainstream value chain. To address the past marginalisation and in light of South Africa's recent political and economic transformation and development agenda, the commercialisation (and associated economic empowerment and mainstreaming) of small-scale and communal farmers has been the mandate of many public and private sector players.

This project was built around the commercialisation of wool production in the communal areas of the Free State Province by developing strategies for concurrent implementation in attempting to manage and overcome various challenges faced by small farmers. The overall objective was to transform communal wool growers' production from an underachieving enterprise to a profitable, sustainable, and renewable venture to enhance the communal wool producers' livelihoods. The project evolved from previous National Wool Growers Association (NWGA, 2022)

DOI: 10.4324/9781003387497-18

research where specific needs were identified to improve the profitability of wool farming in the communal areas of the Eastern Cape of South Africa. The project supports the South African National Treasury's strategy with the extension of the value chain directly to consumers, job creation, and development.

The CARP has built on research for the genetic improvement of wooled sheep in communal farming areas and the introduction of superior wooled rams and improved farming practices. The NWGA has an agreement with the communal farmers whereby they swap an inferior ram in their flock for a superior ram bred by top commercial ram breeders, funded by the national Department of Rural Development and Land Reform. Communal wool farmers own about 26% of the 15 million wool sheep in South Africa. The production of wool from them increased by almost 1,000% from 2000 to 2019, with the highest production of 6.24 million tons of wool produced in 2017 (National Wool Growers Association [NWGA], 2022). The BFAP baseline for the time 2021 to 2030 predicts that the small-scale wool industry can successfully contribute to the export of this highly valued commodity due to well-established public and private partnerships (Bureau for Food and Agricultural Policy [BFAP], 2021). According to the National Wool Growers Association, it is expected that the income of communal wool growers could double to almost R700 million per year (NWGA, 2022). The wool value chain is shown in Figure 15.1.

Research conducted on this project was done by five postgraduate students: two doctoral and three Master's. The studies conducted by the students included the producer, processor, and consumer levels of the value chain. The multi-disciplinary approach of the project resulted in studies completed in the fields of Sustainable Agriculture, Agricultural Economics, Consumer Science, and Animal Science.

The academic collaborators within this project included the Department of Sustainable Food Systems and Development at the University of the Free State, the Central University of Technology, and Glen Agriculture College. The public collaborators were the Department of Agriculture for both the Free State and Western Cape provinces. Private collaborators included the National Wool Growers Association (NWGA) and two large cooperatives in the wool industry, BKB and OVK. The collaboration between private, public, and academic institutions proved fruitful throughout this project. It facilitated the budding of valuable networks, mentorship relationships, and potential for future collaborations within the South African wool industry.

Methodology

hat was needed to set up the project?

Funding for the CARP allowed for installing and developing the relevant infrastructure (such as a small wool processing factory on the university experimental farm). This catalysed further investment by the university in the infrastructure at the experimental farm. A rigorous search was required to identify the correct

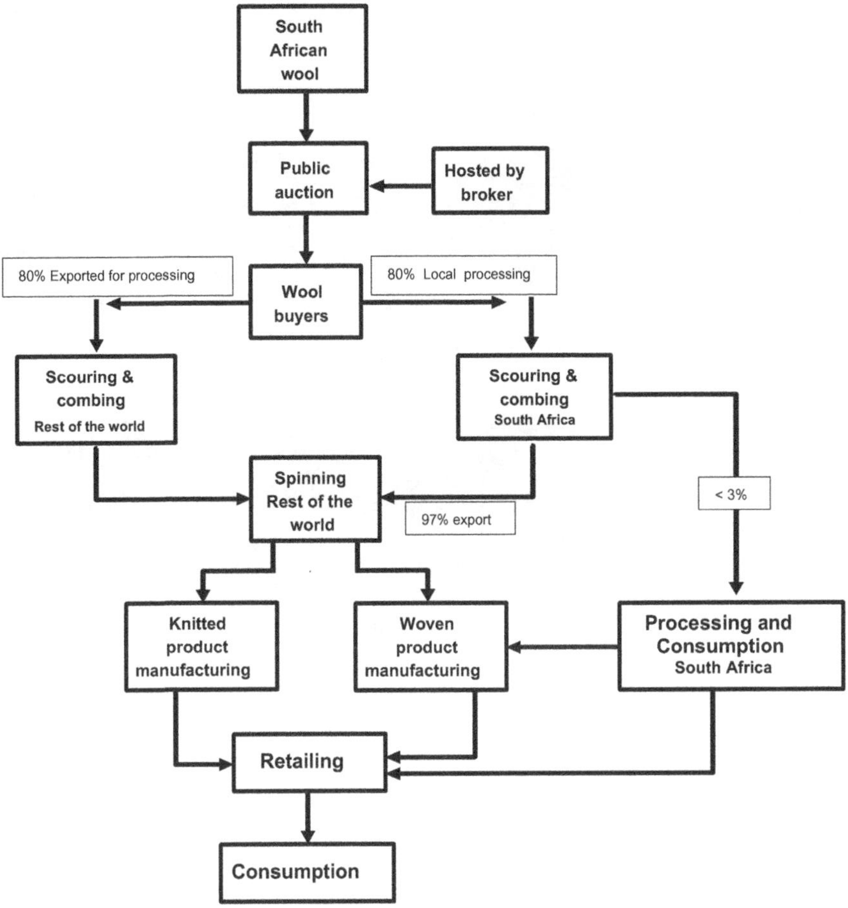

FIGURE 15.1 The wool value chain

candidates and study leaders to conduct all the relevant research and to facilitate training with the students and farmers to achieve the objectives successfully.

How did we go about finding the right partners?

Great efforts went into finding the right partners for the project. Previously built networks were explored and many collaborations initiated. The project was well covered in local media and further collaborations stemmed from this. Partners include, amongst others, from the private sector, industry partners like the National Wool Growers Association, BKB and OVK and Standard Bank, the Provincial Department of Agriculture, and the TVET college. Collaboration was enhanced between departments in the Faculty of Agriculture and with other faculties within the university such as the Faculty for Economic and Management Sciences, and

FIGURE 15.2 Research with a sheep farmer to improve wool yield and profitability

Occupational Therapy. COVID-19 proved to be challenging in terms of the TVET interns that were involved in the project.

Hurdles that arose?

The project began at the end of 2019, and due to the COVID-19 pandemic, in March 2020, South Africa entered the first stage of a national lockdown. University campuses were closed, and all travel and training were halted. The pandemic resulted in project delays as data could not be collected, and training could not be conducted. Due to further restrictions on the number of people per gathering, group training was limited to small groups which hindered access to the community.

Research background

This project is designed to incorporate research and dissemination components and to work in a participatory way between all collaborators. The research component evaluated, tested, and further developed specific technologies and social and economic situations throughout the wool value chain. The lecturers and students from the university worked closely with communal wool growers, commercial farmers, and private industry for training and data collection.

Five studies stemmed from the project and aimed to cover various parts of the value chain. Studies focused on the production level, processing level, and consumer level. The studies collaborated well to provide insight into the entire value chain to better understand farmers' challenges and the potential for improvement at the farm, processing, and market level. The research conducted in the project includes two PhD theses and three Master's theses. Both male and female students had equal opportunities to participate in the project. To facilitate cross-learning and mentorship, the PhD students assisted the Master's students. The five students, university staff, and commercial farmers collectively mentored the TVET interns.

The first doctoral student focused on profiling small-scale wool growers to understand the obstacles that result in poor production and marketing practices while the second PhD candidate conducted research to enhance the livelihoods of small-scale wool producers by developing models that help to improve their efficiency and competitiveness. Both studies concluded that these farmers' marketing and production methods could have been more efficient and were often unprofitable and they proposed interventions to improve yield and returns. The first Master's candidate conducted research within the animal science area and focused on wireworms or *Haemonchus contortus*. This parasitic nematode negatively influences the health and production of sheep. The study evaluated the effectiveness of FAMATCHA™ as a diagnostic tool for small-scale farmers (Besier et al., 2016). The second Master's candidate conducted research within the area of consumer science and evaluated the efficiency of Electrochemically Activated Water (Catholyte) as a sustainable wool scouring detergent and its impact on natural wool dyeing. The third Master's candidate conducted research within the area of agricultural economics and evaluated consumer markets to identify niche markets. The study contributed by identifying those niche markets with demographic and socioeconomic variables used to define target consumers willing to pay the highest premiums for such products.

TABLE 15.1 Students and their research topics

Topic	*Student*	*Degree*	*Research area*
1	Andries Strauss	PhD	Building competitiveness for small-scale wool growers by developing the wool value chain in the Free State province of South Africa
2	Alina Ntsiapane	PhD	Smallholder wool production for improved rural livelihoods in Thaba Nchu and Botshabelo of the Free State province of South Africa - Analysis of wool value adding at small-scale wool farming and its impact on international wool trade of South Africa – Study 5
3	Bonga Madvibi	M.Phil	Prevalence of *Haemonchus Contortus* in small-scale sheep farmers in Central Free State
4	Ketshepileone Shiela Matlhoko	M.Phil	The exploration of ECA as a sustainable alternative for wool scouring
5	Michelle Marais	M.Phil	Identifying a niche market for local, hand-made, and socially responsible wool products in South Africa

Research results summaries

Topic 1

Research into the factors negatively impacting the competitiveness of small-scale wool producers included the lack of basic shearing infrastructure. For example, woolsheds are non-existent for many of the farmers (50.7%) and many do not have equipment like shearing kits. Many of the farmers lack training in wool farming and, instead, learn about wool production from their experience (learning by doing) (64%) and from family members (41.6%). A positive correlation was found between monthly income and owning cattle and pigs. However, the different sheep breeds did not influence the monthly household income. An essential application used for the success of farming operations was a cellphone; however, many farmers did not have access to data/internet (47.1%). There was a significantly low weaning percentage of 47.5%. Most mortalities experienced were due to disease, parasites, theft, and inadequate food due to droughts. The combined factors mentioned above affect these farmers' low productivity and profits and result in insufficient production planning, operation and, marketing (Strauss, 2022).

Topic 2

The study found that most small-scale wool farmers do not have shearing machines or access to storerooms on the farm. Farmers must sell their wool to 'dictators' as they do not have the storage space to keep it. This results in wool that is of poor quality that is sold at a lower price. Overall, these factors all contribute to inefficient wool production. Only a small percentage of the small-scale farmers could be categorised as efficient, as most were not producing at full capacity. Thus, there is much room to improve production. Implementing an assets-based approach in farming can improve the efficiency and sustainability of government projects and enable farmers to be more profitable with their production. Furthermore, this study explores whether value addition could assist in strengthening the international wool trade (Ntsiapane, 2022).

Topic 3

This study looked at animal health and parasites. It showed that the fecal egg count was highly prevalent from January until June. Individual animals experienced heavy worm egg counts of 13,500. Such a heavy worm load is considered severe and fatal if not treated in time. Heavy loads were experienced during May which is regarded as outside the peak period of internal parasitemia. The highest worm egg counts were recorded from January to May. From July to December, low and insignificant counts were recorded. The results clearly show that the use of dewormers during this period is not justified and unnecessary. This is important knowledge to share with farmers to help them lower their costs by focusing dosage in the right months. The study's results confirmed the hypothesis that the sheep are resistant to internal parasites or are resilient to internal parasites due to natural selection (Madvibi, 2022).

Topic 4

Research to reduce the negative impacts of cleaning wool for processing was carried out and the less environmentally damaging wool cleaning with Catholyte (cleaner) was assessed. The results showed that Catholyte removed dirt and vegetable matter from wool and significantly impacted wool colour change. The SEM images showed no damage to the wool fibre and resulted in a wool weight loss of up to 35%, although Catholyte did not eliminate lanolin from the wool fibre. Wool scoured with chemically formulated detergents also significantly impacted wool colour change and resulted in wool weight loss of up to 38%. The results showed that the type of wool scouring agent influenced wool dye absorption. The dyed wool scoured with chemically formulated detergents resulted in low colour strength values and high colour reflectance. In contrast, the dyed wool scoured with Catholyte resulted in high colour strength and low colour reflectance values. Therefore, this study shows that Catholyte has the potential to be used as a scouring agent that might help preserve natural resources and promote a healthy environment through sustainable wool processing (Matlhoko, 2022).

Topic 5

In assessing the potential niche markets for wool handicrafts, this study showed that within the high-income consumer segment of South Africa, 71% of that consumer market prefer locally produced wool products over imported wool products, 55% prefer hand-made over machine-made wool products, and 64% prefer wool products with an indication of social responsibility. These consumer preferences showed that consumers within the South African market value ethical intangible product attributes. This study showed that 52% of high-income consumers are willing to pay a premium of 94% for a local, hand-made, and socially responsible wool blanket, which provides valuable insight for determining the feasibility of the market opportunity for producers and investors in the handicraft wool industry (Sneddon et al., 2012).

Training

Training and support to smallholder wool growers included in-depth training on sheep management, sheep handling, sheep health, and the classing of wool. Sheep shearers were also trained in sheep shearing, while members of the community were trained in the different aspects of value addition to wool by producing a various range of products. Farmers were constantly mentored when they needed assistance. During training initiatives, private partners from industry like NWGA, BKB, and OVK were involved. The connection with these role-players assisted the farmers in developing formal market engagements and readily available information. The linkages and support continue.

Business results

The smallholders were trained and brought together in the platform, improving their returns from wool sales. The reject wool clippings led to small business ventures. A range of different marketing approaches were, and are still, used. Wool products produced on the experimental farm by women from the surrounding community are marketed on Facebook under the UFS Wool Wise page. The products are occasionally sold at farmer's markets in the area. These women are now extending their business to supply products to conferences and the university. Some women who attended training on making wool products by hand have started making and selling their own products outside the project. Two of the postgraduate students worked on a line of jewellery made from felt which they are selling.

The animal science student is currently consulting, on a small business basis, and has become well-known for sharing knowledge and expertise in the surrounding community on animal health and herd management. Some of the students and farmers who have attended training sessions through the project are exploring completing additional training and certifications for shearing sheep and classing wool, as these are marketable skills that will enable them to earn a stable and very good income.

FIGURE 15.3 Training on sheep shearing and wool classification

FIGURE 15.4 Training community women to process and produce felt handicrafts and spin wool

Impact

There was an uptake of innovations in communities. The community largely benefited from this project in two main areas: Firstly, in the profitability and production of wool by small-scale farmers, and secondly, the women in the community acquired skills to produce quality, marketable wool products by hand. Furthermore, wool farmers were trained in better herd management, financial record keeping, wool classing, and wool shearing. These added skills have helped the farmers to improve their productivity and efficiency. The women involved in the project were provided with entrepreneurial opportunities as they learned the skills to produce and sell their products in the future.

The project contributed positively to the effectiveness and employability of the students by providing them with the opportunity to gain valuable practical experience within their fields while completing their studies. Students were enabled to facilitate training sessions with the farmers, where they developed skills in teaching and training. Working with professionals within the industries could further facilitate strong networks and valuable mentorship for these students in the future. In addition, soft skills such as time management and communication were developed during the project.

The project created opportunities for more multi-disciplinary work and collaboration across departments and faculties within the university. The wool project became well-known at the university and a well-referenced example of successful collaboration between multiple departments. The project facilitated five postgraduate studies within four academic departments, allowing collaboration and extended learning.

The project strengthened the university's relations with BTVET and charted a way to create mutual benefits for greater collaboration between post-secondary institutions. The University of the Free State collaborated with Glen Agriculture College and the Central University of Technology. The collaboration built relationships between the institutions as students could learn from each other in a practical environment. Students could travel between institutions and gain exposure to different academic environments and cultures.

Stronger links between the university and communities were created, giving farmers access to the university; this allowed farmers to assist universities in having more demand-driven research and curricula. In addition, farmers could build stronger relationships with each other, forming a more robust network in their communities, which positively impacted their production, marketing, and transaction costs. Valuable information was made accessible to the farmers from the university and provided the farmers with resources such as training guides and working group sessions.

Lastly, the university's social profile was raised as the project became well-known amongst farmers in the area and enabled significant improvements in farmers' livelihoods. Furthermore, the women in the community benefited from processing wool and producing products during the training that had taken place. Students gathered valuable international exposure by entering and winning the IFAMA (International Student Essay Competition) in 2020, making national news in South Africa. They also attended and presented papers at international conferences.

FIGURE 15.5 Displaying the products at international conferences

Conclusion

The CARP has been highly beneficial to all the participants in the project in building valuable networks and establishing sustainable collaboration efforts. These collaboration efforts have provided evidence that capacity-building research projects can mutually benefit all parties who engage and actively participate. The project could also be replicated in the wool industry of another province or country, furthering research and development. The project's success suggests that it is sustainable, replicable, and scalable. One of the main successes of the project is that it received buy-in from key stakeholders, both from the university, and private industry, through financial investment, knowledge transfer, and information centres. Focusing on a given value chain or industry at a time facilitates valuable networking and mentorship relationships which are essential in developing the skills and relationships required to improve and grow the industry sustainably.

Interested readers are referred to the Department Website for further information and a film: www.ufs.ac.za/natagri/departments-and-divisions/Sustainable-Food-Systems-and-Development/research/ruforum-project.

Acknowledgements

This platform would not have been possible without the active support of our collaborators, the farmers and communities, Glen Agricultural College, BKB, OVK, the National Wool Growers Association (NWGA), supervisors Professor Edward Nesumvuni, Professor Henry Jordaan, Dr W. A. Lombard, Dr Jasper Cloete, Dr Jana Vermaas, Mr Leon Kruger and support from Vice-Chancellor Professor Francis Petersen, Professor Danie Vermeulen, Professor Johan van Niekerk, Professor Frikkie Neser, Ms Carien Denner, Mr Johan Barnard, and Ms Doretha Jacobs.

References

Besier R. B., Kahn L. P., Sargison N. D., and Van Wyk J. A. 2016. 'The pathophysiology, ecology and epidemiology of Haemonchus contortus infection in small ruminants'. *Adv. Parasitol.*, 93: 95–143.

Madvibi, B. 2022. 'Prevalence of Haemonchus Contortus in small-scale sheep farmers in Central Free State'. The RUFORUM 18th Annual General Meeting, 12–16 December 2022, Harare International Conference Centre, Zimbabwe.

Matlhoko, K. S. 2022. 'Implementing sustainable wool value chain practices to maximise earnings for wool farmers and the community of the Free State: A case study in the Free State'. The RUFORUM 18th Annual General Meeting, 12–16 December 2022, Harare International Conference Centre, Zimbabwe.

National Wool Growers Association (NWGA). 2022. 'Kommunale boere kan groot vonk wees'. Retrieved on 18 November 2022 from https://www.nwga.co.za/blog-articles/leon-de-beer/communal/kommunale-boere-kan-groot-vonk-weesnbsp/136.

National Wool Growers Association (NWGA). 2022. 'SA wool industry statistics'. Retrieved on 22 June 2022 from https://www.nwga.co.za/.

Ntsiapane, A. 2022. 'Enhancing wool productivity and income for smallholder wool farmers through training'. The RUFORUM 18th Annual General Meeting, 12–16 December 2022, Harare International Conference Centre, Zimbabwe.

Sneddon, J., Lee, J., and Soutar, G. 2012. 'Making sense of consumer's wool apparel preferences'. *Journal of the Textile Institute*, 103(4): 405–415.

Strauss, A. 2022. 'The socio-economic profile of the small-scale/communal wool farmers in the Mangaung district of the Free State Province, South Africa'. A. J. Strauss, University of the Free State, South Africa. The RUFORUM 18th Annual General Meeting, 12–16 December 2022, Harare International Conference Centre, Zimbabwe.

The Bureau for Food and Agricultural Policy (BFAP). 2021. 'BFAP baseline outlook – 2021 to 2030'. Retrieved on 23 December 2022 from https://www.sagis.org.za/BFAP-Baseline-2021.pdf.

16

CLIMATE-SMART AGRICULTURE

Improving dryland crop yields and value addition through university–community partnership in Zimbabwe

Ronald Mandumbu, George Nyamadzawo, Wadzanayi Innocent Nyakudya, Agathar Kamota and Friday N. M. Kubiku

Introduction

Decreasing rainfall has worsened the food security situation of most smallholder farming communities in southern Africa that depend on rain-fed production systems. National agricultural production in Zimbabwe, especially from smallholder farmers, relies on rain-fed agriculture which is vulnerable to climate change and variability (Mandumbu et al., 2021). According to the Zimbabwe National Climate change response strategy of 2016, unless appropriate interventions are made, the negative impacts of climate change and variability on agricultural productivity will stress institutional structures at grassroots level. Communities living in the semi-arid natural farming regions which offer very limited livelihood options, make up 64% of the land area in Zimbabwe (Brazier, 2015). The focus of this CARP was on these semi-arid zones in the Mount Darwin district in the northern part of Zimbabwe and the Buhera South district in the southeastern part of Zimbabwe. These smallholder farming communities have been ravaged by the effects of drought. Marginalised local communities with traditional practices of agriculture are the first ones to face the impacts of climate change. The survival of these communities depends on the effective adaptation of agriculture to climate change. This project was directed at climate-smart agriculture to support small-scale farmers in the semi-arid communal farming areas of Zimbabwe.

Irrigation to increase agricultural productivity is expensive and very limited in scope. In the absence of appropriate infrastructure for capturing water at field scale, up to 50% of the rainfall can be lost from cultivated fields through runoff (Nyamadzawo et al., 2013). Improved in-field water harvesting technologies can conserve moisture that is enough to take the crop through drought stress periods reducing crop failures and improving household food security and livelihoods.

DOI: 10.4324/9781003387497-19

Another adaptation mechanism includes the adoption of more climate change-resilient crops such as *Pennisetum glaucum*, *Eleucine coracana*, *Sesamum indica* and *Vigna unguiculata*. These are called Africa's neglected crops due to their non-coverage in international research and food systems. These drought-tolerant crops have largely been ignored. Typically, they are grown as food sources of last resort, and knowledge of their production, processing and marketing has not been effectively tapped. Furthermore, value addition of these crops has not been much explored. Maize (*Zea mays*) and groundnuts (*Arachis hypogea*) have more research available but were included in some of the CARP interventions in response to a request from the local farmers wanting technologies that could be suitable in their remote and semi-arid areas.

The community action research project (CARP) assessed water harvesting and strengthening crop value chains as an approach to address food insecurity and climate challenges. In-field water harvesting strategies such as tied-contour rain water harvesting systems are methods with the potential to increase the resilience of small farms to climate change. Improved in-field water harvesting technologies can conserve moisture that is enough to take the crop through longer drought stress periods hence crop failures are reduced and household food security is attained.

Marginalised communities (such as the CARP-targeted districts in Mount Darwin, the Zambezi Valley, and Buhera South) with traditional agricultural practices and very limited access to information and markets are vulnerable to climate change. They need to be able to improve yields of their drought-tolerant grains and legumes. And once a surplus becomes more available, they need information on proper grain handling, storage and value addition at community level. These farms are small and remote and marketing costs are high. They need to be supported to aggregate their production and work in groups to market and gain more market power. This will improve small-scale farmers' participation in the national economy.

The farmers in the two semi-arid districts selected for this project were trained in moisture conservation at field scale, design and management of the cropping system and post-harvest handling skills. The CARP gave them better access to information and training for improved productivity as well as value addition training for product development and marketing.

The CARP project brought an approach of shared vision and mutual understanding and trust building between the university and the communities. This approach contrasts with the ivory tower approach which sees the university as the only source of knowledge and innovation whilst the community is at the receiving end (Bowers, 2017). The university is located in a community and should play a role in the welfare and development of that community (Sathorar and Geduld, 2021). In Zimbabwe, this new collaborative approach supports one of the most important new roles given to the university by the central government: to interact with the community in a way that brings social and economic transformation (Education 5.0, Zimbabwe Govt.). Furthermore, the model used by the CARP project was not the charity model but the justice model. Morton (1998) reported that the charity model occurs when resources are given from the university community to another community whereas the justice model recognises that resources are mutual and shared amongst members of these same communities.

The Bindura University CARP focussed on millet (*Eleucine coracana* and *Pennisetum glaucum*), sesame, groundnut and cowpea value chains with a specific focus on production (moisture conservation and cropping system development), post-harvest handling, value addition and product development. The project also used locally available fruits growing wild, in particular, masau (*Ziziphus mauritiania*), to add flavour and value to the products. It was envisaged that after uptake of the water harvesting and integrated pest management systems, production would increase, and farmers will have surplus produce.

The specific objectives of this project were: to promote tied-contour rain water harvesting systems for improving small grains and legume production, capacity building of smallholder farmers in good agricultural practices, and the establishment of three value-addition and product development centres in the two districts.

Small grains value chains in the target district

The participation of small-scale farmers in the small grains in the two CARP-targeted districts has always been limited to the production end of the value chain. These farmers produce the grain (including maize and legumes) and sell it to the middlemen who hire transporters and resell the produce in the urban centres for better prices. There is usually no value addition at community level except for home consumption. The farmers' produce is usually sold at very low prices. The CARP project intervened in the small grains and legumes value chain at production stage and at processing, value addition and product development stages.

Project impacts

Student training

One of the intended impacts of the CARP project was student training. Eleven students successfully graduated with Master of Science degrees (one from the University of Zimbabwe and ten from the Bindura University of Science Education) (Table 16.1). Their study areas covered tied-contour rain water harvesting systems, small grains production, post-harvest handling of grains and product development. They were supervised by researchers from multiple disciplines including crop science, environmental sciences, agricultural engineering, agricultural education, agricultural economics and extension. The crops studied included groundnuts, sesame, sorghum, millets and cowpeas.

The CARP project also supported technical and vocational training college students (TVETs) and the project partnered with Magamba Training College in

TABLE 16.1 Postgraduate students

Name	*Surname*	*Research topic*
Runyararo Evelyne	Motsi	An analysis of the impact of contract farming on the livelihoods of cowpea smallholder farmers in Mt Darwin District, Zimbabwe
Vimbayi Christie	Manyisa	Development of sesame food products to enhance food and nutrition security in Zimbabwe
John	Chinyama	Enhancing food availability from own entitlement through strengthened value addition of sorghum in semi-arid regions
Newman	Choto	Yield and Aflatoxin levels in groundnuts planted on ridges against flat seed beds under dry land farming in Mt Darwin
Andrew C.	Muzanenhamo	The effectiveness of integrated pest management in controlling cotton mealy bug (*Phenacoccus solenopsis*) in sesame (*Sesamum indica)* production under semi-arid conditions: the case of Mt Darwin district, Zimbabwe
Daniel	Machingura	Efficacy of ash, Garlic and chili on Fall armyworm, *Spodeptera Frugiperda* on the management of Maize
Fungisayi Susan	Makuzwa	The impact of Groundnut value chain commodity Associations on women farmers in Zimbabwe, Mt Darwin District Casa study
Merlin Carren	Kandima	The effect of sesame agribusiness (production and marketing) on small holder farmer's household food security
Tapiwa	Zengeza	Tied-contour rain water harvesting in sesame-cereal intercropping systems in semi-arid Zimbabwe
Desire	Marongwe	Developing a Sesame and sorghum-based porridge
Rudo Florence	Mapfeka	Influence of storage facilities on post-harvest losses of small grains in smallholder farming communities of Buhera South and Mt. Darwin Districts, Zimbabwe

Mutare, Zimbabwe. The programmes included a diploma in agriculture and certificates in carpentry, building, welding and garment making. A total of 34 students were supported by the project and they have since graduated with various qualifications. These students came from remote farming areas where they would not have easily been able to access further education. Upgrading their skills made them more marketable for self- (or formal) employment and, also, brought their knowledge back into their remote farming home areas.

The collaborators in the Bindura CARP project were the farming communities of Mount Darwin and Buhera South Districts. The students participated in the demonstration set up of TCRWH systems and their evaluation by farmers. The project also collaborated with the Department of Agriculture Technical and Extension

Services (AGRITEX) where there was interaction with the extension officers who assisted in the implementation of the project. This deepened links with the university and, also, provided greater links between the agency and the communities.

Community university partnerships

The CARP approach provided a platform for university-community partnerships and gave profound experience both to project collaborators and the students. The CARP approach was beneficial to the university as it increased its visibility within the community and enabled them to respond directly to the expressed farmer research needs. The establishment of farmer field schools[1] provided a lasting method for the introduction of new technologies to the communities. The CARP approach also enabled the establishment of a mutually beneficial relationship between the university and the communities. The setting up of the community value-addition centres provided opportunities to improve livelihoods and food security. The project also directly contributed to improving the welfare of youth in the area.

Another objective of the CARP project was capacity building of the lead farmers and extension officers. Across the CARP initiatives, over 5,000 farmers and extension officers were trained on tied-contour rain water harvesting, post-harvest approaches and farming as a business.

Research results

The students engaged with the farmers to assess a range of technologies in their conditions and then worked with them to develop products. The CARP team held discussions with the farmers to determine what their key constraints were and M.Sc

TABLE 16.2 Number of farmers trained on various aspects of the CARP project

Training	*Number of participants*	*Skills imparted*
Tied-contour rain water harvesting systems	4,000	• Making the TCRWH • Merits of TCRTWH and other attributes • Setting up demonstration plots in farmers' fields
Post-harvest crop management	1,000	• Principles of drying • Structures for post-harvest handling of cereals • Post-harvest pests and their management • Processing at farm level
Farming as a business, innovation and product development	500	• Farm records • Equipment inventory records • Profit and loss • Gross margin budget Innovation cycle • Product development and commercialisation

students then carried out research to address their concerns, in the process sharing knowledge and providing demonstrations for rapid uptake. Tied-contour ridges are small earthen ridges, 15 to 20 cm high, with an upslope furrow which accommodates runoff from a catchment strip between the ridges. The results reflected increased moisture retention. Intercropping systems were assessed and combined with integrated pest management to support farmers to improve yields. The use of a biopest mix of ash with a garlic-chilli mix proved to be an effective strategy for addressing the fall armyworm challenge. Farmers were trained using farmer field schools and a women's co-operative was supported with safe and higher-valued groundnut processing and marketing arrangements as well as equipment for processing. These interventions by the project helped farmers get better value for their crops.

Sesame/cereal intercropping under tied-contour rain water harvesting systems showed increased yields because of the higher (0.05) soil moisture content in fields with TCRWH compared to standard contours for both sesame-finger millet and sesame-maize intercrops. The highest moisture content was obtained from a one-metre distance from the contour and moisture became less and less with increasing distance from the contour. Sesame performed better as a sole crop than when intercropped. However, when productivity was measured using land equivalent ratios, the intercrops performed better than monocropping. An increase in sesame densities in sesame-cereal intercropping increased the overall productivity of sesame. The research concluded that when sesame is included at higher densities, the overall productivity is increased.

FIGURE 16.1a Village in Zambezi valley study area

FIGURE 16.1b Sesame farmer and Agritex officer

The tied-contour maintained higher moisture content across the post-planting period (Figure 16.3). Soil water content was higher in the tied contours at 1 metre and 13 metres from the contour although it was the same for both tied and modified standard contours at 7 metres from the contour (Mandumbu et al., 2021).

Groundnut production in Mount Darwin district is affected by poor agronomy hence this study sought to determine the effects of the method of planting and groundnut varieties on aflatoxin occurrence and productivity. One variety, Kasawaya which is a local variety, had the highest ($P<0.05$) mean pod count and a shelling percentage of 71.55%. Guinea fowl variety showed the least ($P<0.05$) leaf spot infection. Both methods (flat and raised bed) of land preparation did not significantly ($P<0.05\%$) affect the aflatoxin prevalence. This reinforced the local farmer perspective of using the locally adapted varieties to optimise yield and to reduce aflatoxin.

An assessment of the factors affecting the use of post-harvest grain structures in small-scale farms in Mount Darwin showed that educational level, farm size, post-harvest training, access to extension services and total harvest increase the intensity of usage of post-harvest structures. Laboratory analysis showed that the highest grain damage ($P<0.05$) and grain loss occurred in the grain bags stored under kitchen conditions whilst the highest germination percentage occurred in the grain stored in a metal tin in the living quarters. The study recommended metal tins and plastic buckets for storage of grain, preferably kept in a granary.

Product development

The University and the Mount Darwin community partnership has increased the productivity and crop yields of small farms. In addition, it has given rise to the development of products including an instant porridge made from pearl millet, sesame and *Ziziphus mauritiana* acting as a biofortifier and a flavourant. This product has passed the prototype stage and is nearing commercialisation and is branded as Dande Insta porridge. The CARP team has also produced Sopese porridge from sorghum and sesame. The team has produced sesame lunch bars (white and black), sesame butter, and sesame oil. *Ziziphus mauritiana* pulp has been used to make a beverage and peanut butter is now produced in the processing centres. The university community partnership has also produced a Rupiza (mixture of roasted cowpeas and peanut butter) for use as a protein rich soup. Preliminary market research and taste testing of all products is promising.

Most of the products are nearing the commercialisation stage. Preliminary market testing indicates acceptability and potential for demand even outside the rural communities. The CARP project bought equipment for product development which included a dehuller and a grinding mill which are housed in the university's innovation hub. Product development and commercialisation led to university-wide collaboration across departments which include the Department of Chemistry for product analysis, Department of Commerce for marketing and commercialisation and Department of Computer Science for social media branding.

The CARP project is in line with the Ministry of Higher and Tertiary Education, Innovation Development and Industrialisation which introduced the Education 5.0 policy in universities in Zimbabwe. The policy emphasises the role of the university in research, teaching, community service, innovation and product development. The policy places innovation and industrialisation at the centre of university functions. The CARP project has helped to raise the visibility of the university and the team members, positioning both as drivers at the forefront of innovation (The Herald, 2022).

The construction of the community value-addition centres also provided equipment and self-employment opportunities to youth and women in rural areas. It is a university-community partnership which indicates mutuality of the two entities. The CARP project provided building materials such as roofing sheets, cement and planks whilst the community provided the builders, labour, water and construction sand. The construction process took the shortest possible time because of the mutuality of the two entities.

Shared university-community responsibility for building the value-addition centres

The CARP project university-community journey for the past four years has led to improvements in yields of up to 40%, ensuring food security for many of the

FIGURE 16.2 Products at prototype stage on display at international conference: (a) displaying products, (b) sesame butter, (c) fortified sorghum, millet and sesame flour

poorly resourced farmers. The project also introduced the concept of community value-addition centres. The CARP project has made it possible for communities to produce products from the raw materials from their farms. The farmers can now make their own porridges, sesame oil and butter.

The CARP project helped to raise the profile of the university with local communities. The team benefited immensely through better collaboration. The project also introduced three farmer field schools through which the performance of the

FIGURE 16.3 Building Kaitano value addition centre

TCRWH system was tightly monitored and challenges were observed. It is a practical way of communicating new ideas to an audience and was effective as part of the CARP approach including the extension services and other researchers and stakeholders.

The CARP project has enabled BUSE to prepare educated, engaged citizens for addressing societal issues. The CARP project fulfils the call for institutions to become more responsive to the society's needs and bears the responsibility to research and share with the community. The work has only just begun and there are now opportunities for much greater collaboration between the university and the communities.

Conclusions

The implementation process also had its own challenges including the COVID-19 disaster and the volatility of the Zimbabwean currency and logistical difficulties. This project has provided the farmers with more climate-smart options for facing the future. Areas for further research along the value chains include product development, improved marketing and the scaling up of farmer field schools and community value-addition centres. The CARP project improved the relations between the university and the communities where the project was being implemented. The development of products has opened opportunities for further collaboration with the communities in sustainable management of biodiversity with an increased focus on wild fruit trees and improved water harvesting and management of crops. The CARP project has opened opportunities for collaboration and bringing back the neglected, more drought-tolerant, crops back into the main food systems for the communities and the country. Given the challenges of climate change, the CARP provided a way for the BUSE researchers to be relevant to their communities and to strengthen climate-smart agricultural value chains in the semi-arid regions of

Zimbabwe. The project generated knowledge, produced products, graduated students, empowered the communities through establishment of community value-addition centres and farmer field schools and trained more than 5,000 lead farmers and extension officers. This project has highlighted the potential for universities to change their communities and contribute directly to national goals.

Note

1 FFS is a farmer participatory approach where an innovation is applied jointly by a group of farmers on a field called the farmer field school model field (van den Berg et al., 2021).

References

Brazier, A. 2015. *Climate change in Zimbabwe: Facts for planners and decision makers*. Konrad Adenauer Stiftung.

Bowers, A. 2017. 'University-community partnerships models: Employing organisational management theories of paradox and strategic contradiction'. *Journal of Higher Education Outreach and Engagement*, 21: 37–65.

Holland, B. A. and Gelmon, S. B. 1998. 'The state of the "engaged campus": What have we learned about building and sustaining university-community partnerships'. *AAHE Bulletin* 51: 3–6.

Mandumbu R., Nyawenze, C., Rugare, J. T., Nyamadzawo, G., Parwada, C., and Tibugari, H. 2021. 'Tied ridges and better cotton breeds for climate change adaptation'. In *African Handbook of Climate Change Adaptation*, ed. W. Leal Filho, N. Oguge, D. Ayal, L. Adeleke, and I. da Silva, 181–195. Springer Nature. DOI: https://doi.org/10.1007/978-3-030-45106-6_23.

Morton, K. 1998. 'The irony of service. Charity, project and social change in service learning'. *Michigan Journal of Community Service Learning*, 2: 19–32.

Nyamadzawo, G., Wuta, M., Nyamangara, J., and Gumbo, D. 2013. 'Opportunities for optimisation of in-field water harvesting to cope with changing climate in semi-arid smallholder farming areas of Zimbabwe'. Springer Plus 2: 100. DOI: https://doi.org/10.1186/2193-1801-2-100.

Sathorar, H. and Geduld, D. 2021. 'A critical approach to university-community partnerships: Reflecting on the diverse realities'. *Educational Research for Social Change*, 10(2): 88–104.

The Herald. 15 October 2022. 'Varsity cashing in on masau fruit'. https://www.herald.co.zw/varsity-cashing-in-on-masau-fruit.

van den Berg, Henk, Jan Willem Ketelaar, Marcel Dicke, and Marjon Fredrix. 2020. 'Is the farmer field school still relevant? Case studies from Malawi and Indonesia'. *NJAS: Wageningen Journal of Life Sciences*, 92(1): 1–13. DOI: https://doi.org/10.1016/j.njas.2020.100329.

Zimbabwe Government. 2021. 'Education 5.0 The modernisation of Zimbabwe through education, science and technology development'. Ministry of Higher and Tertiary Education, Science and Technology Development. Retrieved on February 2023 from https://library.buse.ac.zw/docs/gvt-publications/higher-edu-plan.pdf.

17

ENHANCING COMMUNITY ADAPTATION THROUGH CLIMATE-RESILIENT AGRICULTURE

Encroacher bushes value chains initiative in Otjozondjupa Region, Namibia

Simon Angombe, Cecil Togarepi, John Mupangwa and Hikevali Ndinomuwa

Introduction

The study focused on encroacher bush-based commodities in the Africa Wild Dog Conservancy in Okondjatu[1] a settlement located in a communal area where job opportunities are limited. The CARP baseline survey found that half (49%) of the participants earned less than US$28 per month reflecting family cash income of less than US$2 a day. The region is inhabited by mainly livestock farmers whose livelihoods have been negatively affected by bush encroachment. The encroacher bushes have reduced livestock productivity through thickening of undesirable woody species (especially *Senegalia mellifera* and *Terminalia sericea* species) that have replaced rangeland pasture grasses (de Klerk, 2004). This project was established to assess whether it is possible to turn the encroacher bush problem into an opportunity for economic benefits through value-addition and value-chain development. Bush encroachment has compounded the challenges already faced by livestock producers in the area such as frequent droughts and the effects of climate change. Bush encroachment has affected the livelihoods of the farming community negatively through a reduction in rangeland carrying capacity and hence a reduction in the number of livestock marketed annually (Mlunga and Gschwender, 2016; Gitonga et al., 2020). Furthermore, bush encroachment reduces biodiversity making it important to find a way to reduce these negative impacts to both farming and nature.

Importation of livestock feed in Namibia amounted to about US$231 million, ranking as the eighth most imported product (Nkonya et al., 2016). This is not sustainable, and imports are increasing in frequency as Namibia becomes more susceptible to droughts and climate change. Harvesting thorny bushes and utilising them as livestock feed and other products may contribute to combating bush

DOI: 10.4324/9781003387497-20

encroachment (Hosbein et al., 2017), reclaiming encroached rangelands, improving animal feed security, and contributing to increased livestock productivity. Bush encroachment is viewed as an environmental problem (Shikangalah and Mapani, 2020), because of its threat to land available for farming activities, and the carrying capacity of the rangelands has been reduced by up to two thirds leading to economic losses in the livestock sector (Paschino, 2020).

To ameliorate the encroacher bush problem, methods are being used to clear the bush through mechanical, chemical, and manual means. This provides an emerging opportunity of utilising these harvested encroacher bushes to develop various products. Eliminating encroacher bush enhances regrowth of the palatable grasses and raises the ability to generate income from products such as charcoal, biochar (Sangkhom et al., 2012; Leng, 1990), wood chips for energy production and livestock feed (Abdulrazak et al., 2000; Mokoboki, 2011; Kamati, 2019), as well as using wood chips as mushroom growth substrate (Megersa et al., 2013).

This project investigated the potential of adding value to the bush value chain and focused on encroacher bush-based animal feed production, bush substrates for mushroom production, fencing poles, and biochar production (Figure 17.1). In addition, the project also studied the production economics of fencing poles from encroacher bushes and evaluated the effect of feeding encroacher bush silage on milk yield, composition and quality of milk products in lactating goats and cows. The effect of bush-based feed on growth performance and meat characteristics of goats was also evaluated and an economic analysis of the alternative uses of bush feed was carried out. The project has four MSc and three PhD students (Table 17.1) with support from 12 BSc students and eleven supervisors across multiple disciplines.

This project continues the work done by the Sustainable Management of Namibia's Forested Land (NAFOLA) project. NAFOLA was a five-year project funded by the Global Environmental Facility (GEF) implemented by the Ministry of Agriculture, Water and Land Reform to strengthen and enhance community forests by encouraging communities to manage their natural resources. The CARP established an innovation platform including primarily (1) the University of Namibia, (2) the local TVET institution (Okakarara Vocational Training Centre) and (3) communities from the African Wild Dog Community Forest/Conservancy. Other collaborators were the Ministry of Agriculture, Water and Land Reform, Bush Control and Biomass Utilisation (BCBU), Debushing Advisory Service (DAS), Namibia Biomass Industry Group (N-BiG) and FARM4TRADE (an international livestock data and trade company).

Approach

Establishing a platform and creating linkages with partners at the beginning of the process of developing the project was not easy. Although there were existing relationships that had been established through a team member in a prior project, expectations were not aligned and understanding of the approach was limited. The development of the proposal was in some instances delayed as information was

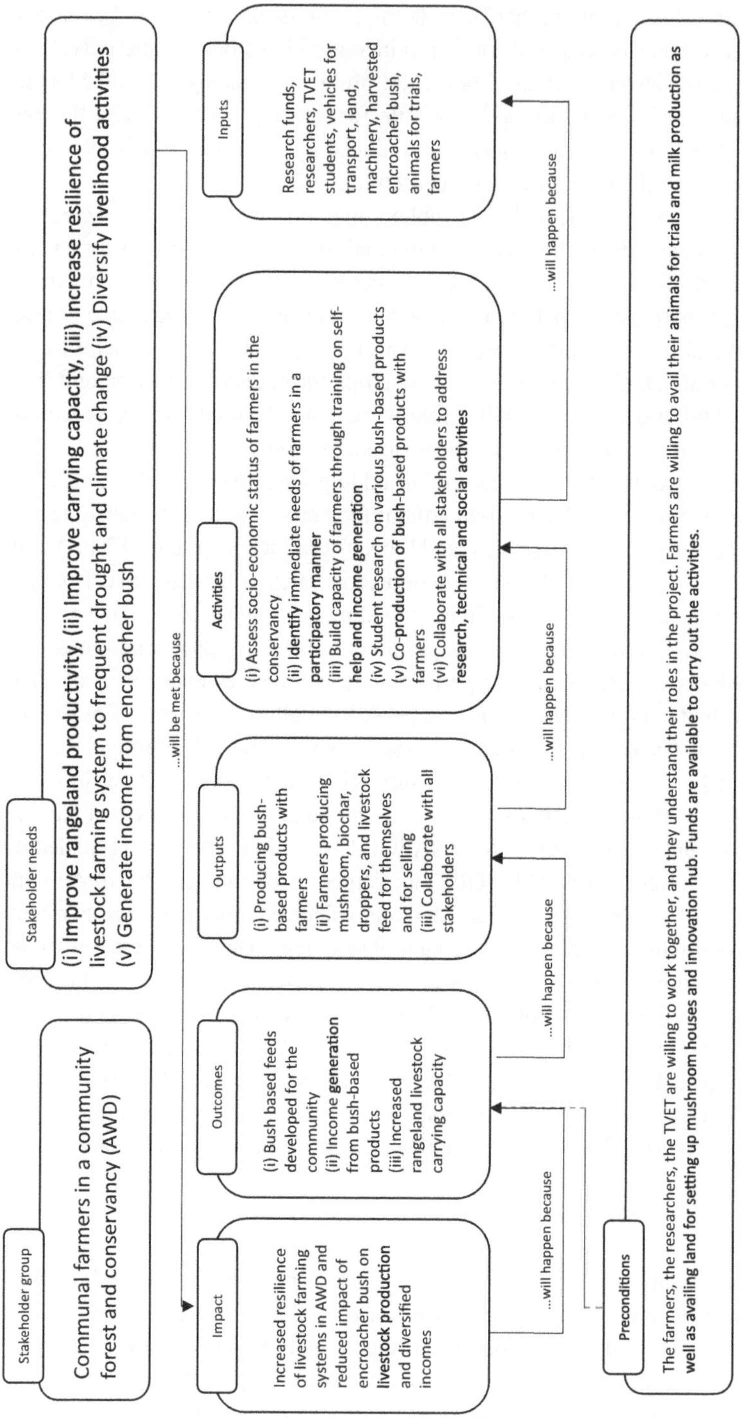

FIGURE 17.1 Theory of change

required from different partners. These partners had not written a research project proposal before and thus a lot of time was required to explain to each partner for them to provide the necessary information. However, existing relationships helped the process. The CARP added value to the earlier project and consolidated their existing operations and diversified some of the activities which the partners were involved with. The community partner (AWD) had existing harvesting and processing equipment for the encroacher bush (see Figure 17.2). Though the communities had such equipment, they lacked the necessary knowledge and skills on feed formulation and fortification and were underutilising the equipment. Furthermore, the community lacked artisans to construct or fabricate the infrastructure that was required to operate and maintain the equipment. This is where the TVET institution was integral to the project through the technical students in the partnership. The project also tapped into the experience of Debushing Advisory Services (DAS) in working with communities and the business development approaches to be used.

(a) (b) (c) (d)

FIGURE 17.2 Bush clearing and milling for bush feed: (a) the large milling machine; (b) towing the machine to farms; (c) small milling machine in operation; (d) cutting branches for the small milling machine

The use of the CARP approach had an immediate impact as it put research into use by those who need the outputs and processes the most. In our case, the community and farmers needed supplementary feeding for their livestock as they were faced with recurring droughts, increasing encroacher bushes and reducing grazing capacity. Thus, alternative feeds were needed. Other options to supplement the feeding of the livestock are too expensive when using available commercial products. As such, the approach dealt with needs assessments and unlocking potential for alternative livelihoods for the farmers that included mushroom cultivation, value addition to milk, and bush feed formulation. The CARP encouraged women and youth to participate in the local economy in these new commodities that the communities were not previously involved in. The approach was focused on inclusivity which seemed not to have been the case previously as only the conservancy leadership seemed to be aware of the opportunities that encroacher bush offered. However, amongst the challenges the project faced initially were that the conservancy leadership had their own expectations of how the CARP project would assist them which were not aligning with the project objectives. As a result, it took longer to agree on the way forward on project activities and how each party would be involved in the project. Thus, although results that are generated from a project through participatory research have an immediate impact, the process which involves initiating and establishing the protocols with a community is cumbersome. The challenges are due to the fact that all stakeholders need to understand their roles first and trust each other as well as dispel any mistrust and misconceptions of expectations.

Transfer of skills was made easier through the CARP approach as the community was involved in the activities: bush harvesting, milling and feed formulation, bush conservation as silage, mash, and pelleting. Bush-based rations were formulated and evaluated on their effect on the growth performance of goats and milk production of indigenous lactating cows in the Okondjatu communal area from the local farmers. The project also investigated the milled bush as a substrate for mushroom cultivation. The students also had first-hand experience working with the communities and researching problems to meet the needs of the communities.

The postgraduate students have worked widely with a range of supervisors and associate supervisors and reported below are the preliminary results. The doctoral students are expected to complete by 2024 and the Master's students in 2023.

Results

Effect of bush-based feeding on Boer goats' growth

The farmers produce Boer goats primarily for meat but also for the milk. In good rainfall years, they do not need to supplement. But in the dry season, and especially after long droughts, supplementary feed is needed to improve goat growth especially goat weaners and other growing animals. The CARP research revealed that

TABLE 17.1 Postgraduate students areas of research

Student name	*Degree*	*Thesis title*
Cecil Togarepi	PhD	Estimating the Economic Value of the Encroacher Bush Bioeconomy in Namibia
Magdalena Kamati	PhD	The effects of feeding encroacher bush silage on milk yield, milk composition and reproductive performance of lactating cows
Nahas Angula Enkono	PhD	The Ecological Impact of Bush Encroachment on the Communal Rangeland Ecosystem of Okondjatu
Portia Murorua	MSc	An evaluation of the effects of different preservation methods on the nutritive value of *senegalia mellifera* bush feed
Kristine Haukongo	MSc	Comparative study on value chain, yield, and nutritional aspect of Namibian commercial mushrooms cultivated on post-harvest encroacher bushes as substrates in Otjozondjupa region
Tashinga Mabambe	MSc	The effect of bush-based feed on milk and milk product's composition and quality from Holstein Friesian cattle at Neudamm campus, Windhoek, Namibia
Joyvin Kanuameva	MSc	Determining the nutrient and anti-nutrient composition and evaluating the effects on goats of feeding *senegalia mellifera* or *terminalia sericea* bush-based feeds

there are no differences in the nutrient content, namely ash, organic matter, crude protein, calcium, and phosphorus content of *S. mellifera* and *T. sericea* rations. The effect of feeding the bush-based rations and the standard grass-Lucerne commercial ration was evaluated in a 90-day goat growth study. The feeding trial using bush-based diets indicated that they are capable of supporting moderate growth rates of 120 g/day in weaned goats and 160 g/day in weaned lambs (Shiningavamwe, 2022). This means, there is no need for the farmers to buy expensive feeds, but rather they can produce their own feed from the encroacher bush with the addition of supplements. Despite the positive response to different encroacher species as roughage sources, the performance was lower than the proposed figures by van der Merwe et al. (2020) for commercial profitable production where producers aim for an average daily gain (ADG) of 300 g/day. However, the growth rates achieved when the goats and lambs were fed bush-based feed enable smallholder farmers to utilise a locally available feed resource whilst improving the rangeland carrying capacity through bush clearing and utilisation.

Effect of silage bush-based feed on milk yield of lactating cows

Livestock farmers in Okondjatu keep both goats and cattle for multiple purposes that include draft power, meat, milk, social functions, and savings. The farmers traditionally milk their lactating cows for home consumption – thus improving household nutrition – and for marketing within the community. The CARP project evaluated the effect of feeding *Senegalia mellifera* encroacher bush silage on the milk yield and milk composition of Okondjatu farmers' indigenous Brahman cross and Nguni cross lactating cows. The animals were managed according to the farmers' practice which involves hand milking the cows in the morning before releasing them for grazing in the paddocks. The bush-based silage quality was improved by the addition of molasses and malted maize flour as fermentable carbohydrate sources (Kamati et al., 2022). The bush-based silage ration was fed at 5 kg/d for each cow. A milk sample of 500 ml was obtained from each cow and transported in a cooler box with ice and a sub-sample of 45 ml of the sample was preserved (with a broad-spectrum Microtabs) until used for laboratory analysis. The milk yield ranged from 1.4 to 1.8 litres per day for cows on natural grazing only while the cows on natural grazing supplemented with the bush-based silage feed had mean yield ranging from 1.5 to 2.5 litres per day. The milk composition of the cows on rangeland grazing was similar to that of the silage-supplemented cows. The protein content was similar between the treatments ranging from 3.0 to 3.1% while fat ranged from 2.4 to 2.7% in control cows and 1.7 to 2.9% in silage-supplemented cows; lactose ranged from 4.2 to 4.6% and 4.4 to 4.8%; solids-not-fat ranged from 8.1 to 8.5% and 8.0 to 8.0% in control and silage-supplemented cows respectively (Kamati, 2022).

Effect of bush-based feed on milk yield and milk products

The CARP project evaluated the effect of feeding *Senegalia mellifera* encroacher bush-based feed on milk yield and quality of milk products in lactating Holstein-Friesian cows. The bush-based silage feed had high-fat content, total solids, and titratable acidity in comparison to the milk from cattle fed with commercial dairy mix feed. The milk from the cattle which were fed the commercial dairy mix feed had high contents of the following parameters such as density, solid-non-fat, lactose, salts, and proteins. Analysis was also done on the yoghurt processed from the milk as a value-added product. The yoghurt of both treatments was similar. The bush-based feed has transformational potential for agriculture within the communities, as it leads to improved feed availability in the short term and improved grazing over time. However, the cost of harvesting and processing the bush into feed may be prohibitive due to the cost of equipment needed. This capital cost adds to land value on privately-owned ranches making such investment more economical.

Preservation of encroacher bush

Converting the biomass from the encroacher bushes into livestock fodder is one best management practice of controlling bush encroachment while increasing the availability of ruminant feeds sustainably without competing with human food production. Milled bush from encroacher bushes has been used by farmers in Okondjatu as a roughage source to sustain their livestock, especially during the dry season and climate-induced drought. During these periods, there is low quantity and quality grazing available for livestock. As a result, UNAM through the CARP project sought to investigate the potential of conserving the encroacher bush during the full leaf growth stage for use during the dry season. The farmers in the Okondjatu area have used milled bush mash. Thus, the intervention of the CARP project incorporated the mash conservation method and other additional preservation strategies that included bush silage making and bush-based pellet production. The addition of molasses, malted barley flour, and Lactobacillus inoculum produced well-fermented and preserved *S. mellifera* bush silage. The silage kept its excellent quality during the 100-day storage period (Murorua, 2022).

FIGURE 17.3a Livestock bush-feed made from milled *S. mellifera*

FIGURE 17.3b Dried and packed

Effect of bush substrate on oyster mushroom production

Amongst the project objectives on bush value addition was using bush substrate to grow mushrooms. The oyster mushroom species (*Pleurotus florida*, *Pleurotus sajor-caju*, *Pleurotus ostreatus*, and *HK35* [hybrid species]) were harvested from three replicates using bush substrate from the following trees: *Senegalia mellifera*, *Terminalia sericea*, *Glavia flava*, *and Combretum collinum*. All four encroaching bushes used as substrate to cultivate various mushroom species have proven to be feasible to grow mushrooms. The mushroom species *Pleurotus florida*, *Pleurotus sajor-caju*, and *Pleurotus ostreatus* performed better compared to the hybrid

FIGURE 17.3c Goats for bush-based feed research

species. They have also shown antioxidant activity and appreciable nutrient and mineral composition, thus can be used as a remedy to illnesses and a supplement to human diets. Therefore, it is essential to promote awareness and train farmers on mushroom cultivation (Haukongo, 2023).

Capacitate communities on mushroom production

Forty-five farmers (16 women) were trained in mushroom production with 25 of them being younger than 36 years old. None of these participants had grown mushrooms before and they were not familiar with eating these cultivated mushrooms. After training, over half indicated they would grow mushrooms for income, 29% for domestic consumption and 7% for both. Mushroom production provides an opportunity to engage youth and diversify incomes dominated by livestock farming activities. After the training, only 2% said they did not feel prepared to produce mushrooms commercially, but the constraint was the capital to establish production.

FIGURE 17.4 Mushroom production using bush-based substrate

Socio-economic analysis and food insecurity of the community

Assessing participant income gave an indication of their economic status. Apart from the study on income, the project also assessed the status of food insecurity which coincided with the result above that most of the communities lived below the poverty line. The food insecurity of the respondents was evaluated using the HFIAS methodology and revealed that 51% of the communities in the study area were severely food insecure (Togarepi and Charamba, 2023). Although the respondents are traditionally livestock farmers, due to recurrent droughts, livestock numbers have decreased, and this is also exacerbated by bush encroachment, making harvesting of the bush even more important as many subsist on livestock.

An Analytical Hierarchy process was carried out to determine the most important commodities produced in the bush value chain. The communities indicated that poles and charcoal were the most suitable bush-based products based on respondents' rankings (Togarepi et al., 2022a). This highlights the potential that the Forestry Act results in a lack of opportunity to produce and market the products. The Act prohibits the commercial use of forest products especially the production of charcoal which has been touted as the "black diamond." Using qualitative analysis and ranking, our analysis showed that droppers were economically and environmentally more suitable, while animal feed has a lot of potential which remains untapped, and charcoal and biochar would not work at the moment because of regulatory limitations of the Forestry Act. The Forestry Act provides for communities to utilise forestry products for housing and housing of animals as well as for energy provision but not for activities such as charcoal making. The use of bush

FIGURE 17.5 Products from bush clearing include poles for construction

substrate for mushroom production did not feature in the hierarchy because the communities had no knowledge of this business venture.

A demand analysis of encroacher bush-based livestock feed reflected that although farmers are interested in bush-based feed, they are mostly only prepared to pay below the benchmark price because they do not yet have confidence in the products and do not value them at the same level as available commercial products (Togarepi et al., 2022b). Demand for bush-based livestock feed will thus benefit from awareness, demonstrations, research evidence and promotion for farmers to be convinced that the nutritional content and quality are comparable to commercial feed.

Impact

This project is still evolving and engaged. There is already a positive impact from the skills transferred to farmers in feed formulation, value addition to milk, mushroom production and the awareness of the value of encroacher bush and its potential economic value. The community is now aware of the various commodities of economic value that can be derived from the encroacher bushes as well as the processes involved in getting permits to harvest bushes. The project is interdisciplinary in nature as it involved biochemists, environmentalists, animal nutritionists, rangeland management specialists, and agricultural economists who forged collaborations between departments and units. The university, TVET institution, and the community have created synergies and laid a foundation and a platform for future collaborations and engagements. Through this project, engagements with the Ministry of Environment, Forestry and Tourism, the Ministry of Agriculture, Water and Land Reform as well as other organisations such as GIZ, DAS Namibia, and N-BIG were enhanced. The project has also opened opportunities for students to explore, research, and formulate products that have value to the communities and the nation at large to solve real problems, such as feed formulation from encroacher bushes, using encroacher bush chips as a growth substrate for mushroom among others.

The findings from this project have implications for policy review as results suggest that rural communities are limited by tenure and by legislation when they want to utilise the bush in comparison with their counterparts based in the private leasehold farming areas. This opens discussions on how best the rural communities can benefit from encroacher bush which is a problem for biodiversity and for livestock productivity by utilising the bush to make products of economic value. These products have been found to be economically viable even without the capital value that reducing the encroacher bushes provides to large-scale private landowners. The CARP project has enhanced community engagements and action research of the university as well as collaborating with other stakeholders and institutions. The participatory approach the project has taken has accelerated the research into use as the communities immediately applied what they learned.

Conclusion

The CARP project gave the authors an opportunity to work with and support the communities. This type of research encourages the community to participate, engage, and take ownership of the activities implemented in the field. The idea was to understand the individual's experiences through collaborative activities and reflections. As such, it connects academic interests and addresses questions that are significant to the community of Okondjatu. Such opportunities enable us to better support rural communities especially on livestock production systems and bush value chain.

The collaboration with the Okakarara Vocational Training Centre was not smooth because of the following challenges: Firstly, change in leadership management at the institution that changed during the project implementation. Such change brought delays as the new management needed to grasp and understand the project. Secondly, the long distance between the two organisations (TVET and communities). There is a 100 km gravel road between the community and the TVET institution. Such a two-hour drive discourages the already strained staff from travelling to the project site. Such a situation prohibits daily engagement with the communities and vocational training centres for project implementation. Thirdly, the TVET institution does not have agriculture programs. The Okakarara Vocational Training Centre was chosen because it was the only TVET institution within a 300 km radius. This was one of the major problems linking the TVET to the CARP and the communities making it difficult to nurture the much-needed synergies between them.

This project had positive impacts on research development and capacity building through the participation of researchers and students. The impact and benefit of the University of Namibia and its students and researchers have been crucial to the community of Okondjatu as they have impacted much-needed knowledge and skills to the communities. This can be seen through the transfer of knowledge to the communities especially in the area of bush-based products. Indeed, communities had a different approach to bush encroachment, but they have now seen an opportunity and benefits that can be derived from the bush value addition. The specific benefits include the availability of feeds for livestock, improved ecosystem services, improved carrying capacity and returns on livestock, and increased productivity of the rangelands. These benefits will improve the standards of living of farming households. Apart from the current result achieved by this project on social, environmental, and economic benefits, there is still a potential to boost innovative research to assist the communities. In future projects that would use the CARP approach, there would be a greater need to create awareness among stakeholders on how the project works, what the expectations and roles of each stakeholder would be, and how the implementation would proceed in order to improve efficiency as well as achieving the intended outcomes while solving the communities' needs for impact.

The future of bush clearing requires a review of the Forestry Act which currently prohibits commercialisation of trees and encroacher bush in communal areas. This project assisted communal farmers and the conservancy to apply for the necessary permits. Farmers in the private leasehold commercial area are able to commercialise products from bush clearing without permits and, importantly, they can also capitalise on the improved pastures resulting from bush clearing in the value of the land which makes bush thinning even more economically viable. Bush-clearing results in better pastures, livestock production, and biodiversity. Further research can consider the externalities that would benefit the environment and the communal farmers if initiatives would invest in subsidising bush clearing. This CARP has shown that it is feasible to utilise cleared bush profitably.

Acknowledgements

This platform would not have been possible without the active support of our collaborators in the Ministry of Agriculture and Ministry of Environment, Forestry and Tourism, the Okakarara Vocational Training Centre, the African Wild Dogs Conservancy and the farmers, the N-Big Industry Association and Omega Consultancy, the ZERI (Zero Emissions Research Initiative) and the research supervisors: Professor David Uchezuba, Dr Thinah Moyo, Professor Sartorius von Back, Dr Maria Shipandeni, Dr Absalom Kahumba, Dr Lydia Horn, Dr Tjiurutue Muvari, Dr Stephen Barrion, and Mrs Marjory Kandjou.

Note

1 Okondjatu is in the Okakarara constituency in the Otjozondjupa region of north-east Windhoek, Namibia.

References

Abdulrazak, S. A., Fujihara, T., Ondiek, J. K., and Ørskov, E. R. 2000. 'Nutritive evaluation of some Acacia tree leaves from Kenya'. *Anim. Feed Sci. Technol.*, 85(1–2): 89–98. DOI: https://doi.org/10.1016/S0377-8401(00)00133-4.

DECOSA., D. C. 2015. 'Concept for sustainable labour-based de-bushing in Namibia'. Development Consultants for Southern Africa. Retrieved from https://www.l2b.co.za/Tender/Concept-for-Sustainable-Labour-Based-De-Bushing-in/545809.

de Klerk, J. 2004. 'Bush encroachment in Namibia: Report on Phase 1 of the Bush Encroachment Research, Monitoring, and Management Project'. In *Ministry of Environment and Tourism* (Issue 1). http://the-eis.com/elibrary/sites/default/files/downloads/literature/Bush%20Encoachment%20in%20Namibia_deKlerk2004_1.pdf.

Fara, K. 2001. 'How natural are "natural disasters"? Vulnerability to drought of communal farmers in Southern Namibia'. *Risk Management*, 3(3): 47–63. DOI: https://doi.org/10.1057/palgrave.rm.8240093.

Gitonga, Z., Visser, M. and Mulwa, C. 2020. 'Can climate information salvage livelihoods in arid and semiarid lands? An evaluation of access, use and impact in Namibia'. *World Development Perspectives*, 20, 100239.

Haukongo, K. N. 2023. 'Comparative study on the quality of Namibian commercial oyster mushrooms cultivated on encroacher bushes as substrates'. Master's thesis. University of Namibia.

Inthapanya, Sangkhom and Preston, T. R. 2009. 'Effect of supplementation with sweet potato root and paddy rice on growth performance of local rabbits fed water spinach (*Ipomoea aquatica*) and paper mulberry (*Broussonetia papyrifera*) as basal diets'. *Livestock Research for Rural Development*, 21(10). https://www.researchgate.net/publication/290529646.

Kamati, M. I. 2019. 'Identification, quantification and nutritional characterization of agro-industrial by-products in Namibia as potential supplements of the bush-based feeds'. Master of Science (Animal Science). University of Namibia. Retrieved from http://hdl.handle.net/11070/2653.

Kamati, M. I. 2022. 'The effect of additives on the fermentation and chemical composition of *Senegallia mellifera* (Black Thorn) bush silage'. Paper presented at the 18th RUFORUM Annual General Meeting and Scientific Conference, 9–16 December 2022, Harare, Zimbabwe.

Kamati, M. I., Shipandeni, M. N. T., Mupangwa, J, and Lutaaya, E. 2022. 'The effects of additives on fermentation and nutrient composition of *Senegalia mellifera* (Black thorn) bush silage'. Paper presented at the 57th Annual Congress of the Grassland Society of Southern Africa (GSSA), 25–29 July 2022, Aldam Holiday Resort & Conference Centre, Ventersburg, Free State, South Africa.

Leng, R. A. 1990. 'Factors affecting the utilization of 'poor quality' forages by ruminants particularly under tropical conditions'. *Nutrition Research Reviews*, 3: 277–303.

Megersa, M., Asfaw, Z, Kelbessa, E., Beyene, A., and Woldeab, B. 2013. 'An ethnobotanical study of medicinal plants in Wayu Tuka District, East Welega Zone of Oromia Regional State, West Ethiopia'. *Journal of Ethnobiological and Ethnomedicine*, 9(1). DOI: 10.1186/1746–4269-9–68.

Mlunga, L. and Gschwender, F. 2016. 'Bush encroachment, de-bushing, and energy production in Namibia'. In *Perspectives on Energy Security and Renewable Energies in Sub-Saharan Africa*, ed. Oliver C. Ruppel and Bernd Althusmann, 291–296. Macmillan Education Namibia.

Mokoboki, H. K. 2011. 'Effect of species within season on techniques used to measure nutritive value and anti-nutritional factors in browse tree leaves'. *Life Sci. J.*, 8(S2): 112–119. http://www.lifesciencesite.com/lsj/life0802s/17_6246life0802s_112_119.pdf.

Mturi, F. A. 1991. 'The feeding ecology and behavior of the red colobus monkey *(Colobus badius kirkii)*'. PhD dissertation. University of Dar es Salaam, Dar es Salaam, Tanzania.

Mturi, F. A. 1993. 'Ecology of the Zanzibar red colobus monkey, *Colobus badius kirkii* (Gray 1968) in comparison with other red colobines'. In *Biogeography and Ecology of the Rain Forests of Eastern Africa*, ed. J. C. Lovett and S. K. Wasser, 243–266. Cambridge University Press, Cambridge.

Murorua, P. 2022. 'The effects of molasses and malted barley with or without lactic acid bacteria (LAB) inoculant on the pH, nutritional composition, and *in vitro* digestibility of *Senegallia mellifera* silage'. Paper presented at the 18th RUFORUM Annual General Meeting and Scientific Conference, 9–16 December 2022, Harare, Zimbabwe.

Nkonya, E., Mirzabaev, A., and von Braun, J. 2016. *Economics of Land Degradation and Improvement – A Global Assessment for Sustainable Development.* Springer. http://link.springer.com/book/10.1007%2F978-3-319-19168-3.

Paschino, P., Stocco, G., Dettori, M., Pazzola, M., Marongiu, M. K., Pilo, C. E., Cipolat-Gotet, C., and Vacca, G. M. 2020. 'Characterisation of milk composition coagulation properties and cheese making ability of goats reared in extensive farms'. *Dairy Science*, 103(7). DOI: https://doi.org/10.3168/jds.2019-17805.

Sangkhom, I., Preston, T. R, Khang, D. N., and Leng, R. A. 2012: 'Effect of potassium nitrate and urea as fermentable nitrogen sources on growth performance and methane emissions in local "Yellow" cattle fed lime (Ca(OH)2) treated rice straw supplemented with fresh cassava foliage'. *Livestock Research for Rural Development*, 24(2): Article #27. Retrieved from LRRD, 21 August 2023. http://www.lrrd.org/lrrd24/2/sang24027.htm.

Shiningavamwe, K. L. 2022. 'The nutritional content and the effect of feeding four invader Namibian bush species on growth, methane production and meat quality of Damara sheep'. PhD thesis. University of Namibia, Namibia.

Steinfeld, H., Gerber, P., Wassenaar, T., Castel, V., Rosales, M., and de Haan, C. 2006. 'Livestock's Long Shadow: Environmental Issues and Options'. Rome: Food and Agriculture Organization of the United Nations.

Togarepi, C. and Charamba, V. 2023. 'Food insecurity experience in Otjozondjupa region among livestock farmers, Experience of AWD conservancy'. (Manuscript submitted March 2023 to *Journal of Agriculture and Food Research.*

Togarepi, C., Katshenye, N., and Muzanima, J. 2022a. 'Determination of optimal encroacher bush commodities: An application of analytical hierarchy process in Namibia'. Paper presented at the 18th RUFORUM Annual General Meeting and Scientific Conference, 9–16 December 2022, Harare, Zimbabwe.

Togarepi, C., Shiimi, D., and Shivolo-Useb, S. 2022b. 'An assessment of willingness to pay for encroacher bush-based animal feed in Namibia'. Paper presented at the 18th RUFORUM Annual General Meeting and Scientific Conference, 9–16 December 2022, Harare, Zimbabwe.

van der Merwe, D.A, Brand, T.S., and Hoffman, L.C. 2020. 'Precision finishing of South African lambs in feedlots: A review'. *Tropical Animal Health and Production*, 52: 2769–2786. https://doi.org/10.1007/s11250-020-02282-x.

18

THE EMPOWERMENT OF WOMEN AND YOUTH

Enhancing climate-smart productivity of vegetables and key livestock value chains, Sudan

Muna Mohamed Elhag, Tafaol E. Ali, Lubna M. Musa, Igbal E. I. Mohamed, Mutasim A. M. Elagab and Ali E. Eljack

Introduction

In Sudan agriculture, including livestock, is the hub of economic activities and the source for rural livelihoods. The selected area is semi-arid with variable rainfall that is being further affected by climate change (Osman and Ali, 2021). Small-scale producers in the rural areas of Sudan lack knowledge about modern agriculture technologies and marketing and have poor access to finance. They have low incomes, a low standard of living, and a high unemployment rate exacerbated by youth migration from rural to urban areas. Many rural people have bad food habits due to the lack of nutritional diversity in their diet, in addition to the problems of low income (Hasan, 2022). This project focuses on how the university can make a difference by working with rural women and youth to assist them to make climate-smart decisions, adapt to climate change, and improve their family food security and living standards.

The universities and higher education institutions can play a pivotal role in addressing the UN Millennium and Sustainable Development Goals and the African Agenda 2063. Through innovative teaching, research, and outreach approaches, universities can actively participate in realizing the economic and social development in society in general and for rural communities in particular (AbdelRahaman et al., 2015). Women are the backbone of domestic work and household economy and are primarily responsible for the production, processing, storage, and marketing of agricultural products. This reality is often ignored, overlooked, or forgotten in putting together women empowerment programs. Rural families are the cornerstone of local community development. Women and youth, in particular, have the potential to play a greater role in generating income through different agricultural activities in their local communities. Women play an integral part in agricultural

DOI: 10.4324/9781003387497-21

production as subsistence farmers, cash crop growers, food processors, and livestock owners among other roles. Therefore, empowerment of rural women provides them with an opportunity to influence local decisions so that they more closely reflect their preferences and needs (World Bank, 2015). This could also generate indirect positive impacts such as decreasing poverty and boosting communities' well-being, given that there is a strong link between poverty reduction and gender equality (World Bank, 2011). The focus of this project is to empower women and youth through strengthening their vegetable and small livestock production value chains.

This project focused on:

- Improving the household income and nutritional status in the eastern part of Gezira state, Sudan.
- Strengthening the capacities of women and youth in small scale gardens production (backyard gardens and small livestock production).
- Developing knowledge and communication products for effective dissemination of technologies, innovations, and management practices in small-scale gardens.
- Facilitating university staff to work with a target group for adoption of the processing and market-driven technological packages to improve their impact on household welfare.
- Developing a business incubator and women's cooperative society.

The project identified vegetables and livestock with relatively short production cycles, high and quick returns, good nutrient values, and a low requirement for land or capital investments. This included vegetables such as carrots, onions, eggplant, pumpkin, watercress, and radish. It also included the rearing of goats and poultry. Quail birds were found to be very suitable to the study area being more nutritious, with shorter production cycles, and more tolerant to diseases.

In this project, ten Master's students engaged directly with rural women and youth to implement and assess the impact of the introduction of small-scale gardens with good agricultural practices on improved living standards and nutritional status for children under five in *Alhiglig* village, Gezira state, Sudan. The project provided financial, training, and technical support to rural women and youth with a focus on building their own businesses after the project. Also, two students assessed the impact of providing inputs and better marketing of the vegetables and livestock production through the establishment of a cooperative unit for women.

Approach and methods

Alhiglig village is located between latitude 13.36 south, 15.29 north, and longitude 32.25 west, 34.18 east (Osman, 2017), the eastern part of Gezira state, Sudan. Rainfall is the most important single determining factor for agriculture production, annual vegetative cover, land use and thus human occupation. In the area, the temperature is high all year round. The rainfall shows a substantial variation in incidence, amount, time received, and annual distribution (Elagib and Elhag, 2012). The population of the village is about 150 families; the average size of a household is ten persons. The target groups in this project are women and youth, training students, upgrading the teaching capabilities of Gezira University staff, and strengthening interaction with rural communities.

The activities of the project are organized into six work packages:

- WP1: Assessment of the economic and environmental potential of the small-scale farms
- A baseline survey was carried out to assess the economic situation of women in the area and the existing practices, food cultures, and resources.
- WP2: Selection of agronomically suitable vegetables and identification of available areas for vegetable production, selection of participants and establishment of gardens, and irrigation.
- WP3: Identification of types and number of livestock available in the study area and strengthening their value chain.
- WP4: Capacity building on backyard gardens, water harvesting, composting, rations formulation, and poultry rearing management (training modules and M.Sc. students' research work).
- WP5: Establishment of a women's cooperative unit.
- WP6: Assessment of farmers' perceptions and adoption of the newly introduced technical packages and their impact on household income and food security.

Results and discussion

Experiential learning and postgraduate students

The total number of the M.Sc. intake was twelve students constituting of three males and nine females (see Table 18.1), with supervisors drawn from two government research and development agencies with continuous mentoring and coaching from the project team members. The students really benefited from working with the farmers and also by working together with students and supervisors from other disciplines.

Training and financial support for agricultural production

Training sessions were carried out throughout the project period. Women were supported to grow vegetables in backyard gardens, rearing goats, and poultry as well as receiving training on how to market these products through direct sale or through the cooperative unit that was established with help and direct guidance by one of the students whose research was to assess the economic impact of the project. In addition, women were trained on water harvesting techniques and

TABLE 18.1 Student names and thesis titles

Student name	*Thesis title/specialization*
Alnzeer Fathi Omer Elsiddik	Improve Productivity of Indigenous Chicken Ecotype of Rural Household via Management Intervention in *Alhiglig* Village, Gezira State, Sudan (Animal Production)
Esraa Ahmed Yossef Mohamed	Performance of Lactating Nubian Goats Fed Sorghum Straw and Different Diets, Gezira State, Sudan (Goat Milk Production)
Tasnim Hunin Abdalwhab Mohamed	Performance and Carcass Characteristics of Culled Nubian Male Kids Fed Pigeon Pea (*Cajanus cajan* L.) Haulm and Different Concentrate Levels, Gezira State, Sudan (Goat Meat Production)
Asged Babiker Osman Abaker	Effects of Sowing Method on Plant Density, Yield and Quality of Carrot (*Daucus carota* L.) (Vegetable Production)
Emtnan DawAlbait AlhajAhmed	Evaluation of the Performance of Four Sweet Potato (*Ipomoea batatas* L.) Clones, Khartoum State, Sudan (Vegetable Production)
Mutwakil Ibrahim Norelgalil Alkhder	Evaluation of Composting Techniques and Their Effect on Azolla Production Compared to Other Fertilizers, Gezira State, Sudan (Soil Fertility)
Hala Abdalrhim Abdalla Mohammed	Impact of Empowerment Rural Women Programs on living Standards in *Alhiglig* Village, Gezira State, Sudan (Extension and Training)
Tafaol Elsadig Ali Hasan	Impact of Agricultural Production Subsidy on livelihood of Home Farm Growers, Alhiglig Village, Eastern Gezira State, Sudan (Agricultural Economic)
Esraa Mergani Osman Ali	Hydrological Estimations to Calculate Rain-water Harvesting Potentiality Using GIS and Remote Sensing: The Case Semi-arid Area, Gezira State, Sudan (Water Management)
Yousif Hamed Dldom Gomaa	Design and Evaluation of a Low-cost Drip Irrigation System for Smallholder Farmers (Water Management)
Samah ALtayep Ali ALsakhy	Effect of Nutritional Education and Nutritional Practices for Mothers of Under- Five in *Alhiglig* Area, Gezira State – Sudan Food safety and Hygiene (Nutrition)

application for family drip irrigation as tools for water conservation. The training also included food processing, storage, marketing, how to benefit from animal residues to produce organic fertilizer (compost processing), and other practices that suit their needs to adapt to climate change to sustain their food production. Besides these trainings, other training sessions included: supplementary irrigation, land preparation, cropping pattern, and cultural practices. The good practices consisted of using good vegetable seeds, optimum sowing date, suitable crop density, soil fertility management, weed control, pest and disease control, water conservation measures, field set up, and planting methods. Feeds were formulated by adding value to goats' feeds by giving molasses and straw to provide goats with suitable feed including goat kid fattening. Moreover, one student studied the effect of good management practices to improve local breeds of poultry production and, with his supervisor, developed a training module on quail production and conducted many training sessions in the village.

The women were provided with some inputs (seeds, agriculture tools, organic fertilizer) and connected to markets through the project. The results indicate that most of the women in the village could finance their agricultural activities and gain additional income if they received appropriate support and training in income-generating activities. There was a significant difference in the monthly income and savings between the women who participated in the project and the non-participants (P-value < 0.05). There was a significant difference in the food diversity between the targeted group and the non-participants (P-value = 0.000). There was a significant difference in the employment opportunities provision between the training recipients and the non-participants (P-value = 0.039). Women who participated in the project were able to significantly increase their contribution in education expenses for their children (Ali, 2022).

Project Impacts

Below are the key components of the project outcomes after two years of implementation.

- Improved small farm productivity:
  - Family drip irrigation system installed (5 units) for vegetable production
  - Compost production to enhance vegetable production by using crop and goat residues
  - Local chicken breeds (20 households, 10 chickens) improved productivity through new management practices. Development of training module: Improvement of local poultry breeds production including quail
  - Goat productions (milk and meat) goats' kids fattening and increase milk by using different rations formulation, use a different level of concentrate with pigeon pea, Sorghum, groundnut, and molasses as source of energy
  - 25 families with plots of at least $2 \times 2m^2$ producing vegetables such as okra, carrot, onion, pumpkin, jews-mallow, purslane, garden rocket, pepper, and tomato for their own consumption and they are selling surplus. The women involved in this activity reported that they were now completely self-sufficient in vegetables with some left over for sale
- The unemployed graduate's skills in agricultural activities were retooled through training and learning from community participation in the project.
- The project contributed towards the fulfilment of the University of Gezira philosophy which aims at servicing rural communities; the students and supervisors engaged directly with rural communities, some students for research processes stayed at the village for three to six months and this gave them experience for deep learning and positive interaction with rural communities. Established

FIGURE 18.1a Growing vegetables

strong links between the university and communities – giving farmers access to the university more broadly and assisting universities to have active research sites for engagement in the rural areas.

- Eleven M.Sc. students graduated and were able to develop their careers and improved their employability. One received a PhD scholarship to Indonesia, three

FIGURE 18.1b Harvesting vegetables

were recruited as full-time or part-time lecturers, and four received small grants to deepen their community engagement or start businesses.

- *Alhiglig* Women Cooperative Society established with 90 women was registered at the local authority. The youth were engaged through support for market linkages using social media.
- For sustainability of the vegetable production a modified greenhouse was established with the cooperative by using material available at the nearby market. This greenhouse will be used to produce seedlings for sale and for members for some vegetables

Challenges

Below are the key challenges the project was facing after two years of implementation.

- COVID-19 pandemic to the project (university lockdown for more than nine months).
- Political situation (strike and demonstration every week for most of the period since 2019.
- High inflation rate (US$1=44.5 SDG in 2018; US$1=575SDG in 2022).

FIGURE 18.2a Selling vegetables

FIGURE 18.2b Displaying vegetables

FIGURE 18.3 Women's cooperative greenhouse seedlings

- Difficulties stretching project finance allocated for capacity building in the value chain, facilitating the establishment of a community-based enterprise-business incubation, and supporting engagement with the private sector.
- Traditions persist, such as the social sanction that means it is not possible to sell vegetables to friends or neighbours.
- It also required much time to accord with tradition on taking tea and polite talk at each visit before substantive issues could be discussed.
- Much of the Principal Investigator time was absorbed by team creation and building relations with community.
- Managing conflict between the different actors during the project implementation.
- The students lack adequate communication skills to deal with the rural women involved training and learning time.

Conclusion

The women and youth in the project were connected and better able to communicate with governmental and non-governmental agents through the project. Communication was strengthened between the village and the university, ARC, and other local authorities. It developed the understanding of their culture, needs, and knowledge by faculty and students from engaging so closely with the target groups through the project approach.

Recommendations

Based on the findings of the study, the following recommendations are essential to overcome different constraints that hinder agricultural development and the role of women and youth in rural economies. It was clear that more accessible extension services and training in developing household projects and small enterprise management are essential. There needs to be support and investment to improve extension services, training, and access to microfinance to empower local women and retain more youth in the villages. The role of the university in supporting these efforts could be enhanced if they actively engaged in projects and attachments of their students to work with rural communities.

References

AbdelRahaman, Ahmed M., Samira M. Abdalla, and Isam O. Fadol. 2015. 'The Role of Rural Women Schools (RWSs) on Increasing Income, Improving Nutrition and Food Security of Rural Families, Gezira State, Sudan'. *Agric. Sci. Digest.*, 35(2): DOI: 10.5958/0976-0547.2015.00036.1.

Ali, T. E. 2022. 'Impact of Agricultural Production Subsidy on Livelihood of Home Farm Growers, Alhiglig Village, Eastern Gezira State, Sudan'. Unpublished thesis. University of Gezira, Sudan.

Elagib, Nadir Ahmed and Muna M. Elhag. 2012. 'Major Climate Indicators of Ongoing Drought in Sudan'. *Journal of Hydrology*, 409(3–4): 612–625. DOI: https://doi.org/10.1016/j.jhydrol.2011.08.047.

Hasan, T. H. 2022. 'Impact of Agricultural Production Subsidy on Livelihood of Home Farm Growers, *Alhiglig* Village, Eastern Gezira State, Sudan'. M.Sc. thesis. University of Gezira, Sudan.

Osman, A. K. and Ali, M. A. 2021. 'Sudan - Land, Climate, Energy, Agriculture and Development: A study in the Sudano-Sahel Initiative for Regional Development, Jobs, and Food Security'. Working paper 203, ZEF Working Paper Series, ISSN 1864–6638, University of Bonn, Germany.

Osman, M. A. A. 2017. 'Climate Change Adaptation through Soil and Water Management in Semi Arid Environment: A Case of Smallholder Farmers from Butana Region, Sudan'. Working paper 203, ZEF Working Paper Series, ISSN 1864–6638, University of Bonn, Germany.

World Bank. 2011. 'The World Development Report 2012: Gender Equality and Development'. Washington, DC: World Bank.

World Bank. 2015. 'World Bank Group Gender Strategy (FY16–23): Gender Equality, Poverty Reduction, and Inclusive Growth'. Washington, DC: World Bank.

19

STRENGTHENING THE BEEKEEPING VALUE CHAIN IN WESTERN KORDOFAN STATE, SUDAN

Ahmed Idris, Ali Mohamed, Asma Hamad, Ruhama Abdallh, Osman Khmis and Hana Mohammed

Introduction

Honeybee colonies in Sudan are found in different nesting sites and in different types of hives. Wild honeybees *Apis mellifera* (L.) are widely spread in forests. They establish their nests in holes in trees and fallen logs, holes in white ant mounds, rocks roofs and similar places and the honey is also harvested from these wild bees. Domesticated honeybees inhabit either Lang troth hives or traditional hives. The majority of the traditional hives are of elongated wicker or hollowed log type. These hives are fastened to the branches of trees during the swarming seasons. The Sudanese bees are classified into three sub-clusters; the smallest bee of Sudan *Apis mellifera yemenitica*, Ruttner, the medium bee *Apis mellifera sudanensis*, Rashad, and the largest bee *Apis mellifera bandasii*. The smallest honeybee that is found in Sudan is *Apis florea* which was introduced to Sudan in 1985 (El-Sarrag et al., 1992).

There are an estimated 200,000 honeybee hives in Sudan and some 50,000 beekeepers. Almost all beekeepers use only traditional beekeeping methods. Sudan is an agricultural country with growth in horticultural development and thus the already high demand for honey beekeeping and pollination has increased. The low productivity is further reduced by disease, pests and predators despite the hygienic behaviour of Sudanese honeybee colonies (Mogbel El-Niweiri, 2004). This project also studied the sources and applied methods to reduce disease.

This CARP complemented the Sudan government's plans to improve productivity and renewable natural resource use. There are many beekeeping projects that were initiated by large companies which failed because they either did not work within the community or, among those that did, did not effectively introduce modern beekeeping. This action research programme succeeded because it worked

DOI: 10.4324/9781003387497-22

together with local communities training both male and female beekeepers and involving other partners along the value chain.

Approach

This action research programme was designed to strengthen, sustain and contribute to boosting food security development and enabling vulnerable families, particularly women (Women from East Darfur), to improve their livelihoods. The activities were carried out in communities in the vicinity of *Alfulla*, West Kordofan State. This was achieved by supporting the establishment of apiaries and capacity building of beekeepers to empower actors, training graduates of the agricultural sector and vulnerable pastoral families. The CARP focused on capacity development of traditional beekeepers to improve honey production.

The university collaborated with the local Ministry of Agriculture, the Agriculture Research Corporation and leaders of beekeepers. This was done through an administration farm, workshop and group discussions in the field to transfer modern techniques of beekeeping and new concepts to promote beekeeping. The local communities, some government agents and the migrants coming from Darfur were included in the training. The focus of extension and training activities was in West Kordofan State to improve the capacity of traditional beekeepers, especially women.

A range of experimental work on honey sourced from different areas in Sudan and other countries was also carried out by post-graduate students, see Table 19.1. Their research included assessing the effects of honeybee pollination on sunflower seed and yield quality. Two treatments, honeybee-pollinated and open-pollinated sunflower plants were assigned in an RCBD with three replicates used. Another study aimed to identify some chemical properties of six samples of bee honey obtained from different sources of tree such as *Zizyphus pinachristi* (Sidir), *Acacia seyal* (Talih), *Helianth uannuus* (Sunflower), *Ceratoniasiligue* (Cruob) and two samples purchased from Elobied and Alfolalocal markets, Sudan. The samples

TABLE 19.1 M.Sc graduates and thesis titles

Name	*Research topic*
Asma Hamad Ahmed Hamad	Impact of Honeybees (*Apis mellifera*) on Sunflower in the Field and their Pollination Effect on Yield Attributes and Seed Quality
Hana Abdalla Adam Mohammed	Some Physico-Chemical Properties of Bee Honey from Different Sources, Sudan
Osman Musa Mohammed Khamis	Microbial Investigation and Adulteration Test for Bee Honey
Rahama Gafar Mohammed Abdallah	Some Physico-Chemical and Microbial Properties of Bees Royal Jelly Samples from Different Regions, Sudan

were collected from the study areas to investigate chemical properties, moisture content, ash content, total acidity, total sugar, reducing sugar, fructose, glucose, sucrose, ascorbic acid and mineral concentration. The results showed that honeybee pollination improved sunflower seed viability in terms of germination percentage. Also, honeybee pollination significantly improved plant seedling growth including seedling plant height. From the results and findings of the investigation, it can be concluded that the bee honey collected from different floral origins is significantly varied in chemical composition and nutritive value.

Other studies were carried out to estimate the physico-chemical properties of honey, to estimate the microbial load, to determine the inhibition zone of microbes for honey and to identify some types of pathogenic bacteria. Five samples of honey were collected from Central Darfur, South Kordofan and South Darfur, Central Africa and South Sudan to analyse refractive index, pH, viscosity, moisture, ash, total acidity, hydroxy methyl furfural, total sugars, reducing sugars, glucose, fructose, sucrose, total count of bacteria, yeasts, mould and the inhibition zone of microbes by *Bacillus subtilis*, *Staphylococcus aureus,* and *coliform bacteria*. There was also a study to measure some physico-chemical and microbial properties of three samples of Royal Jelly collected from three regions, *Al Salam* University farm at West Kordofan, *Ustaz Faisal* farm at *Umrwaba*, North Kordofan, and *Mamlakat alnahal* supermarket at Khartoum. The microbial properties assessment included the total viable count of bacteria, mould and yeast count and inhibition zone of two bacterial strains. The results from these studies were used to inform the ongoing research of the Ministry of Agriculture and the CARP implementation.

The core objective of the project was to engage with the communities to increase beekeeping and thus improve sustainable livelihoods, biodiversity and crop production. Effective community engagement was important to be successful in promoting adoption to improve the productivity of beekeeping and change traditional honey production to use the new and improved approaches in West Kordofan State. The platform employed different community engagement strategies to encourage support for local farm businesses and bring other partners into this work. Promoting honey production in the local community was seen as a good way to elevate local food security and improve farm incomes. The community engagement included farmers and migrants from different villages in the study area.

Impact

The project envisaged benefiting beekeeping value chain actors. It changed the concept of traditional beekeepers by using modern beehives and protective clothes compared with the traditional methods in Western Sudan. The project strengthened beekeeping, trained traditional beekeepers and provided employment for the migrated women from Darfur. The project adopted group training on modern honey production with modern hives. Five farmer associations were used to reach 200 people who have had their income increased by adding honey production to their farms. These farmers will also share the new ideas with others in their groups, so many more are expected to be impacted. At the end of the project, 80% of beekeepers were using modern methods and the production of honey from the area has increased by 50%.

In addition, the project developed packaging and linked beekeepers into a value chain to supply urban markets. The use of innovative value-addition processes and access to markets has increased the value of honey. It is likely that the expanded beekeeping has also positively impacted crop production through additional pollination. In addition to the farmers, the CARP has given birth to five beekeepers associations and a micro agribusiness company that has begun to investigate the potential for honey productivity from the trees that are located in the hilly areas.

This CARP benefited Peace University by skilling 12 master's students. It strengthened the university and raised its visibility with both the communities and also local, national and international agencies, especially with different government departments. The outreach programme was developed in partnership with stakeholders. It included regular sharing of lessons with stakeholders (including policy and agribusiness actors), production and dissemination of project outputs thus strengthening relationships. In addition, it contributed to strengthening the Peace University website and the D-space contents with publications.

One of the most successful outputs of this action was that the CARP triggered additional support and external interest to support the engagement of the university with the farmers. International agencies including IFAD and FAO adopted the technologies and coordinated their own activities with the CARP to include honey production in their field activities and introduced these new ideas to the farmers.

References

El-Niweiri, Mogbel. 2004. 'Survey of the pests and diseases of honeybees in Sudan'. M.Sc. thesis. University of Khartoum, Sudan.

El-Sarrag, M. S. A., Saeed, A. A., and Hussien, M. A. 1992. 'Morphometrical study on Sudanese honeybees'. *J. of Agric. Sci.*, 1: 99–100. King Saud University.

20

IMPROVEMENT OF INDIGENOUS COPING STRATEGIES IN FAMINE-STRICKEN DARFUR, SUDAN

Elhadi A. I. Elkhalil, H. A. H. Osman, A. E. M. Elzein, E. M. Mohammed and E. E. A. Ahmed

Introduction

A famine is a widespread scarcity of food, caused by several factors including war, inflation, crop failure, population imbalance, or government policies. Famine is usually accompanied or followed by regional malnutrition, starvation, epidemics, and increased mortality (Dyer and Falsani, 2019). Darfur is one of Sudan's most suffering food-insecure regions. Darfur has been hit by several famines; one of the most serious was as a result of drought in the years 1979–1984. The main causes of famine include war, failure of rain, a malfunction of traditional agricultural plantings, and the use of land for export crops instead of food for the country (Chris, 2018). The gradual desertification process is a long-term problem of survival in the most affected areas of North Darfur. This has resulted in virtually non-existent agricultural production in North Darfur and low production in South Darfur.

The overall objective of this project was to identify and promote indigenous famine survival strategies based on native foods, plants, crops, and fruits in Darfur states using both local knowledge and advanced biotechnological approaches. This was achieved using a participatory approach through multi-disciplinary field surveys carried out to identify the foods and understand the socioeconomic characteristics of famine-stricken people in Darfur states. Primary data were collected to document and perform in-depth botanical and ecological studies of wild edible plants, fruits, grass, and roots as famine foods consumed at times of famine before international relief reaches the area of famine. Traditional habits of consumption of famine foods were assessed to determine availability, nutritional and social acceptability, and people's willingness to adopt proposed changes. Knowledge of the nutritive value of studied famine foods was determined, suitable indigenous fermentation processes were recommended, and storage techniques and the nutritive value of the famine foods were improved.

DOI: 10.4324/9781003387497-23

Approach

The scientific steering committee of the project was composed of five staff members from three departments. The committee selected 12 students (nine women) to carry out the research project. Their research topics were complementary and were designed to identify and promote indigenous famine survival strategies and to research applicable and acceptable solutions using indigenous and advanced knowledge, scientific methodologies, and technological innovations. The research topics covered: (i) Understanding the socio-economic characteristics of the population of the study area and the indigenous practices, attitudes, and survival strategies of the families and households to cope with famines, (ii) Production of adequate quantities of traditional famine foods by identifying, selecting, and using efficient microbial strains to fast-track the fermentation process, (iii) Improvement of the quality, nutritive value and safety of the famine foods through proper fermentation and compatible additives, (iv) Optimization of the storage conditions and packaging systems at the village level for long shelf-life and preservation of the indigenous fermented or unfermented foods (studying the effect of drying and storage technologies and containers, packaging system, and storage temperature), (v) Assuring of high safety aspects of these traditionally prepared fermented or unfermented foods after long periods of storage time as well as (vi) Identification and exploration of new famine foods (untapped sources of famine foods) that are available but not yet been properly investigated and utilized (e.g. root and tuber crops, wild rice and grass, and non-wood forest products (Tasnim, 2021).

The various studies were undertaken by post-graduate students working with communities in Darfur, Sudan.

Summary of key research results

Field surveys were carried out to identify and understand locally used famine foods (Kamal et al., 2022; Marouf et al., 2022). This was followed by nutritive value and chemical composition analysis of the collected samples determining key nutritive values and anti-nutritional factors as well as microorganisms in the fermented foods (Marouf, 2022). The results showed that famine foods were found to be highly acceptable within Darfurian communities and among all age groups (Mohammed et al., 2022). Ambachi (*Dioscorea hispida*), Corape (*Dactyloctenium aegyptiacum*), wild rice (*Oryza barthii*), Cassava (*Manihot esculenta*), *Azanza garckeana*, and *Boscia senegalensis* were fairly plentiful in Darfur states and grow naturally in the wild. All respondents were knowledgeable about their appearance, growth cycle, usage, and storage processes. Virtually all households (99%) consume foods derived from those famine crops, particularly during famine periods, and 66% were familiar with the cooking process. These plants showed good potential for food and fodder (Brika et al., 2023; Ahmed et al., 2022). Most (75%) households consumed fermented foods including Damsora, Fendo, Furundu, and Khemiss Tweira during

TABLE 20.1 Student names and thesis titles

Student name	*Thesis title*
Alaa Emadeldeen Souliman Mohammed	Improvement of Furundu (Fermented *Hibiscus sabdariffa* L. seeds) by Addition of Sesame Cake as Famine food in North Darfur State, Sudan
Misa Ali Almahdi Ahmed	Microbiological and Biochemical Evaluation of Fendo (Fermented Locusts) as Famine Food in Northern Darfur State, Sudan
Farida Abd Elkarim Elkhair Marouf	Enhancement of Nutritive Value and Keeping Quality of Khemiss-tweira (Fermented pearl Millet) as Famine Food in Darfur, Sudan
Ahmed Mohammed Saeed Balola	Microbiological and Biochemical characterization of improved Damsora fermented Famine Food derived from Sorghum flour in Darfur, Sudan
Abdah Abdalla Mohammed Ahmed	Characterization and Assessment of Corape (*Dactyloctenium aegyptiacum*) Plant as Famine Substitute Food in Northern Darfur State, Sudan
Misa Mohamed Brika Ismail	Evaluation of Wild Rice (*Oryza barthii*) Plant as Famine Food in South Darfur State, Sudan
Amera Khater Abdelrasoul Adam	Evaluation of cassava (*Manihot esculenta*) as potential famine-food in Darfur
Marwa Kamal Aldien Sayed	Evaluation and Enhancement Grown Ambachi (*Dioscorea hispida*) in food security during famines in Darfur State
Tasnim Salah Alhassan Bashir	Nutritive Value of Two Indigenous Tree Species Products and Their Role in Famine Relief in Darfur Region, Sudan
Mhasin Ahmed Ibrahim	Household coping strategies toward famine food in kutum locality north Darfur state, Sudan
Yahia Elhag Yagoub Ahmed	Enhancing household's famine coping strategies in East Darfur state, Sudan
Sheikheldin Bakhit Sheikheldin Ahmed	Enhancement of the Role of Orphan Crop Lablab Bean (*Lablab purpureus*) as famine Food in Darfur State, Sudan

famine time. These are prepared by germination, fermentation, baking, drying, and adding sugar and salt – except Fendo which is derived from locusts and was not germinated. The fermented foods could be stored for more than a year without reported deterioration in food value and quality. Over 90% of respondents found them highly acceptable (Ahmed, 2022).

The laboratory analysis showed that all famine foods contain relatively high levels of carbohydrate, starch, protein, vitamin C, fat, and ash contents. In fermented foods, protein content was higher due to fermentation and the role of microorganisms. Famine foods were rich in mineral contents (Na, K, Ca, Mg, Fe, P, and Zn) (mg/100g). The shortcomings of fermented foods include rancidity and nutrient deficiency in some minerals such as Ca, F, and I, which could affect their properties as famine food (Balola, 2022). This could be overcome by supplementation with

(a)

(b)

FIGURE 20.1 (a) Students in the field with farmers; (b) prepared cassava meals

the relatively abundant *Ziziphus spina-christi* L. flour, *Adansonia* (baobab), and *Hibiscus sabdariffa* L. calyx extract that improved the nutritive value and maintained the qualities of fermented foods. These additives are rich in minerals and contain antioxidant compounds. In addition, they are growing wild and are easily obtained.

FIGURE 20.2 (a) Stone grinder "Mohraka"; (b) Corape (*Dactyloctenium aegyptiacum*); (c) Wild rice (*Oryza barthii*), and other wild forest foods: (d) Hanza (Boscia senegalensis); the Azanza (*Azanza garckeana*)

Impacts

The most important impacts of this research project could be summarized as follows:

(i) A variety of nutritious, safe, and long-term storable famine-food products was identified, improved and promoted. A total of ten products indigenous to Darfur either had their nutritive value improved through additives, their packaging and cottage-industry potential developed, or their shelf-life and storage ability prolonged. This has important implications for supporting communities to be more self-reliant in famine-prone areas.

(ii) Technical capacity building and good quality training of 12 MSc students was provided for sustainable promotion, industrialization, commercialization, and usage of the developed and improved famine-food products.

(iii) The university developed relationships with the community and other stakeholders along the value chain. the platform developed will provide opportunities to strengthen the relevance of the university to the local community. These efforts now need to be broadened and deepened to secure the future of people living in famine-stricken areas.

The platform can now be used and extended to include aid agencies, government, and the private sector which could work with the university to develop cottage industries and support the commercialization and the availability and uptake of famine foods. This will help to improve household food security and incomes. Further research and investment are needed to ensure sustainable practices for collection and harvesting and to improve further on packaging, marketing, and awareness raising. The platform needs to be strengthened and scaled out to commercialize some of the products and broaden access to the technologies across at-risk communities. This project indicates that there is considerable potential for good returns on investment both in terms of commercialization and reduction of food aid needs in famine-stricken communities. Sharing the results generated on these famine foods will also benefit other countries and communities in Sudan and in Africa.

The grant has opened avenues of collaboration between the university and other universities in Darfur such as Alfashir and Zalengi universities which were actively involved with the students supported by the grant during their field work in Darfur.

Sudan is an emerging market in terms of food products, especially packaged food. Some dried food packed items, such as fast-produced porridge, have already been produced and they have met with good success. There is still room for more such products. Famine foods could fill a gap, especially for students and workers. Famine foods studied in this research are high-energy fast foods that are considered indigenous traditional Sudanese foods. One such famine food that could be promoted is Damsora. Damsora may become a popular snack similar to the instant noodles (indomie) widely consumed by young people in Sudan. The new products are not only useful in famine but could be used by boarding students, campers, as well as for people in informal mining areas who need easy-to-prepare, nutritious food. The concept of this business is to make famine foods popular and available for the majority of the population.

In the community of women food vendors, we engaged with 17 small-scale food vendors which enhanced the nutritive value of some fermented food as well as improved the shelf-life of products, and modern packaging. In turn, they relayed their

(a)

(b)

FIGURE 20.3 (a) Processed packaged foods; (b) commercialized famine foods

experience to others and hence spread the technologies. The most important benefit for the participants was raising awareness. The survey results showed that previously most of the famine foods studied were either little known or their value was underestimated. The research outcome has created a rising local interest in these foods. It was found, for instance, that some of the grains of some of these famine foods were stored for long periods of time (ten years in some instances) without being utilized in spite of food shortages. Some of the targeted populations did not seem to realize their value and had lost touch with their uses.

Conclusion

Twelve MSc. students acquired knowledge and skills through training courses and workshops and improved their qualifications. Working in interdisciplinary teams in the field and with other agencies has strengthened the capacity of the graduates for employment and opened avenues to establish small businesses. It has been a great opportunity for them as they will be eligible now to join teaching staff in the more than 25 government universities and numerous private universities all of which are understaffed. The Faculty of Agriculture, University of Khartoum, has greatly benefited from the CARP grant through the opportunity for faculty and students to work with communities outside the university to strengthen their curricula and inform their research agenda. The results generated can now be used to link the university to aid agencies and government policymakers in an effort to provide support and reduce the vulnerability of Darfur communities at risk of famine.

Upon completion of the project, it is clear the need for scaling up and sustainability of the projects. The data collected and results arrived at must be further verified by practical application in other communities affected by famine. It should be noted that both security and famine aspects are continuously deteriorating in Darfur. This has been recently exacerbated by the global food crisis resulting from the war in Ukraine.

The selected famine foods in this research project, either the untapped plants or the fermented and packaged foods, could play a vital role in alleviating hunger during famine time with their high nutritive value, long shelf-life, and reasonable price. Furthermore, this research project could be implemented in other regions in Sudan, especially drought areas in eastern Sudan. Improved visibility and approaches to involved agencies through the platform established will be part of the sustainability plan for this CARP.

References

Ahmed, A. A. M., Osman, H. A. H., Elkhalil, E. A. I., Ahmed, E. E. A., Elzein, A. E. M. and Mohamed, E. M. 2022. 'Value of Corape Plant as Famine Food in Darfur State, Sudan'. *African Crop Science Journal*, 30(s1): 67–76. DOI: https://doi.org/10.4314/acsj.v30is1.6S.

Ahmed, M. A. A. 2022. 'Microbiological and Biochemical Evaluation of Fendo (Fermented Locusts) as Famine Food in North Darfur State, Sudan'. M.Sc. thesis. Department of Agricultural Biotechnology, University of Khartoum, Sudan.

Balola, A. M. S. 2022. 'Microbiological and Biochemical Characterization of Improved Damsora Fermented Famine Food Derived from Sorghum Flour in Darfur, Sudan'. M.Sc. thesis. Department of Agricultural Biotechnology, University of Khartoum, Sudan.

Bashir, T. S. A. 2021. 'Nutritive Value of Two Indigenous Tree Species Products and Their Role in Famine Relief in Darfur Region, Sudan'. M.Sc. thesis. Department of Agricultural Biotechnology, University of Khartoum, Sudan.

Brika, M. M., Osman, H. A. H., Elkhalil, E. A. I., Ahmed, E. E. A., Elzein, A. E. M., and Mohamed, E. M. 2023. 'Socioeconomic Evaluation of Wild Rice as Famine Food Crop in Southern Darfur State, Sudan'. *African Crop Science Journal*, 31(1), 113–119. DOI: 10.4314/acsj.v31i1.9.

Chris, H. 2018. 'What Is Famine, and How Can We Stop It'. Oxfam America. https://www.oxfamamerica.org/explore/stories/what-is-famine-and-how-can-we-stop-it/.

Dyer, J. E. and Falsani, C. 2019. *The End of Hunger: Renewed Hope for Feeding the World.* United States: InterVarsity Press.

Kamal, M., Elzein, A. E. M., Mohamed, E. M., Elkhalil, E. A. I., Osman, H. A. H., and Ahmed, E. E. A. 2022. 'Indigenous Knowledge-based Evaluation of Ambachi Tuberous Plant as a Future Famine Food Crop for Darfur State, Sudan'. *African Crop Science Journal*, 30(s1): 53–66. DOI: https://doi.org/10.4314/acsj.v30is1.5S.

Marouf, F. A. E. 2022. 'Enhancement of Nutritive Value and Keeping Quality of Khemiss-tweira (Fermented Millet) as Famine Food in Darfur, Sudan'. M.Sc. thesis. Department of Agricultural Biotechnology, University of Khartoum, Sudan.

Marouf, F. A. E., Elkhalil, E. A. I., Mohamed, E. M., Osman, H. A. H., Elzein, A. E. M., and Ahmed, E. E. A. 2022. 'Insights in "Khemiss-Tweira" as A Famine Pearl Millet Coping Strategy in Darfur State, Sudan'. *African Crop Science Journal*, 30(s1): 45–51. https://doi.org/10.4314/acsj.v30is1.4S.

Mohammed, A. E. S., Elkhalil, E. A. I., Mohamed, E. M., Osman, H. A. H., Elzein, A. E. M., and Ahmed, E. E. A. 2022. 'Socio-demographic Relationships and Management Regimes on Use of Frundu as Fermented Famine Food in Urban Northern Darfur in Sudan'. *African Crop Science Journal*, 30(s1): 37–43. DOI: https://doi.org/10.4314/acsj.v30is1.3S.

21

SCALING UP AFRICAN BAOBAB FOOD PRODUCTS VALUATION THROUGH ENHANCEMENT OF THEIR SAFETY AND VALUE CHAINS FOR FOOD AND NUTRITIONAL SECURITY IN BENIN, WEST-AFRICA

Achille E. Assogbadjo, A. Hounkpèvi, K. V. Salako, Flora J. Chadaré, R. Idohou, Rodrigue C. Gbèdomon and R. Glèlè Kakaï

Introduction

Micronutrient deficiencies are major threats to maternal and child well-being in Africa including Benin. In Benin for example, there are high anaemia rates due to iron deficiency (78.1% in children under 5 years and 61.3% in women of reproductive age, INSAE & Macro International Inc. 2007). With such a situation, improvement of food and nutrition security becomes a must. To efficiently address these issues, wild edible plants and their derived foods are an excellent source to explore. These resources are often rich in micronutrients and can be used in food-to-food fortification. If well structured, their value chains can strategically contribute to improving farmer revenues and hence improve their livelihoods. Among them, is the African baobab, *Adansonia digitata* L., one of the most remarkable trees of the world (Gebauer et al., 2016). In many countries in West Africa including Benin, the leaves and fruit pulp are important ingredients in sauces, porridges, and beverages (Chadaré et al., 2009; De Caluwé et al., 2009). Its importance in human nutrition is well established and this is driving the development of a growing local, and increasingly global, market.

At the same time, the over-utilization of wild baobab saplings has been affecting natural regeneration. Indeed, the species is facing a risk of extinction in the wild where it is threatened because of high pruning for food, bush fire (which destroys seedlings), livestock grazing, and lack of natural regeneration. The BAOCHAIN project was established to enhance the potential for baobab to address food security in the relatively arid regions where they occur. This project addresses issues across the baobab value chain by (i) developing sustainable plant propagation

DOI: 10.4324/9781003387497-24

techniques for leaves and fruit production within farmer fields, making it possible to expand access to this nutritious plant while reducing pressure on wild supplies and conserving biodiversity, and (ii) improving food processing techniques to ensure micronutrients availability and increasing safe baobab-related products for the communities.

The baobab value chain (VC) has been disregarded in many national state agendas. In this context and based on outcomes of earlier research activities on baobab (Assogbadjo et al., 2005; 2008; Chadaré et al., 2014; Hounsou-Dindin et al., 2016), a more comprehensive innovation platform was established to consider the full VC. This was implemented from March 2018.[1] It is a multi-actor participatory action initiative that developed a platform to support a sustainable and operative baobab value chain in Benin for its most valued products (leaves and pulp) for human nutrition.

The overall goal of the project is to combine both participatory research and capacity building activities to establish a long-lasting valorization scheme for the African baobab at a national scale in Benin. Specifically, it aims to (i) diagnose baobab leaves and fruit pulp value chains; (ii) organize actors involved in baobab leaves and fruit pulp value chains (VC) into a platform and build networks to ensure long-lasting connections; (iii) develop market-driven technological packages for baobab leaves and fruit pulp and derived products to encourage baobab cultivation to supply the market with safe, good quality baobab-derived products; (iv) establish pilot incubators for baobab leaves and fruit pulp valorization in the project areas; (v) scale-up added-value novel technologies at country level; and (vi) develop an advocacy plan for better integration of baobab products in food and nutrition security strategies and to support their agribusiness potential at a national level.

The conceptual framework guiding this research responded to a need to improve nutrition, conserve biodiversity, and to support the diversification and income levels of local people. It was designed to strengthen the Baobab VC contributing to meeting food and nutrition security at a national level while at the same time supporting biodiversity conservation and reaching the sustainable development goals of a hunger-free world. The research actions covered two main domains: (i) sustainable propagation of the species to improve its leaves and pulp availability, and (ii) improved approaches for food and human nutrition with baobab-related products. The capacity of 35 students (two PhD, five MSc, eight BSc and 20 TVET students) was developed in these two domains under the supervision of ten university researchers and ten TVET teachers. The Community Action Research Platform involved various institutions including two national universities, two Technical and Vocational Education Training colleges (TVET), five non-university-based research institutions and four private sector partners.

In addition, projects carried out by eight BSc and 20 TVET students contributed to the development of good agricultural practices for cultivating baobab and also to the development of marketable food products. Students competed for small awards to better establish their enterprises.[2]

TABLE 21.1 Post-graduate student research

Name	*Level*	*Topic*
Mechak Gbaguidi	PhD	Post-harvest handling and processing of market-oriented baobab pulp, leaf powder and their derived products across their value chain in Benin
Marius Affonféré	PhD	Utilization of local resources to alleviate iron deficiencies among children aged 6–59 months in Benin: case of *Adansonia digitata* L., *Moringa oleifera* Lam. and *Cochlospermum tinctorium* A. Rich
Mariette Agbohessou	MSc	Comparative analysis of the performances of grafting and aerial marcotting to propagate baobab
Vanessa Idohou	MSc	Development of "ready to cook" seasoned baobab leaf sauce powder
Kevin Fassinou	MSc	Food formulation using baobab (*Adansonia digitata*) fruit pulp and African bush mango (*Irvingia gabonensis)* fruit pulp
Karl S. Agbodossindji	MSc	Use of backslopping in "Mutchayan" production, a local food based of baobab pulp
Maurice Pacôme Solevo*	MSc	Exploitation of baobab (*Adansonia digitata* L.) leaves: ethnobotany, local perceptions, and technical aspects of its cultivation in the context of climate change

Note: *National University of Agriculture. All others at the University of Abomey-Calavi, Benin

Methodological approach

The project was prepared and implemented such that it shifted from traditional conservation and poverty alleviation approaches to the use of a business-oriented approach with a pro-poor growth strategy that supports biodiversity. This business-oriented approach implies the development of business solutions for sustainably benefiting the poor while reducing the impact on wild baobab. It consists of developing conditions and enhancing capacities and abilities of the poor and disadvantaged stakeholders to participate in, contribute to, and benefit from the transformation of the baobab value chains. The project, a multi-disciplinary and multi-agency team, established a participatory research-for-action innovation platform.

The research carried out within the project was the focus of two PhD students (with a background in agronomy but specializing their doctoral work in food science). Their research work was in synergy with five MSc students, eight BSc students, and 20 TVET students. The student research covered all the work packages which were carried out under the supervision of a collaborative research team involving researchers from university and TVET partner institutions. Outputs of these research works were discussed with other partners and with organizations of farmers. Eight scientific articles resulted from the research and the methodology is well presented in six of these articles (Gbaguidi et al., 2020, 2022; Affonféré et al., 2021a, 2021b; Agbohessou et al., 2020; Idohou et al., 2018). The activities of the project started after an inception meeting that allowed all partners to get together to discuss methodological adjustments and to adapt the implementation approach for synergy and efficiency.

Results

The research part of the project relied mainly on the activities of the doctoral and master's students which included work on food sciences and on propagation of baobab (see Table 21.1). The research activities in food sciences resulted in the development of five technologies: Seasoned baobab leaf powder; Complementary Food Supplement (CFSs); *Mutchayan*, a fermented cereal dough enriched with baobab pulp; Maize flour enriched with baobab pulp; Pasteurized baobab nectar drink. The research activities on the propagation of baobab enabled the development of three technologies: baobab leaves production within 45 days and, drawing on earlier research, final slot grafting and side veneer grafting to support earlier fruiting.

Mechak Gbaguidi's research aimed to improve the quality of baobab pulp and leaf products focusing on (i) documentation of the baobab pulp and leaf properties, (ii) fermentation methods of cereal-based doughs, (iii) optimization of baobab leaves oven drying, (iv) characterization of baobab pulp foods in the main Benin cities, (v) assessment of microbial and physico-chemical changes during *Mutchayan*[3] spontaneous fermentation and, (vi) evaluation of back-slopping fermentation effect on *Mutchayan* quality. Major findings (Gbaguidi et al., 2020) from his research showed that baobab pulp and leaf-derived products were the most preferred products in the principal cities of Benin (Cotonou, Abomey-Calavi, Porto-Novo, and Parakou), and that they are rich in vitamins (especially C, B1, B2,), minerals (especially Ca, Fe,), and antioxidant compounds (in particular procyanidins, and tiliroside). Several benefits of fermentation are reported: enhancement of minerals bioavailability, bioactive compounds production, and product safety improvement. Optimal drying of baobab leaves is at 45°C for 23.5 hours. The presence of Enterobacteriaceae was reported in baobab pulp samples found at markets which suggested unhygienic processing conditions. *Mutchayan* quality was improved by the back-slopping fermentation applied. Spontaneously fermented, *Mutchayan* was characterized by the rise of titratable acidity (from 5.7 to 6.8 g LA/100g dry weight) and antioxidant capacity while dry matter, Brix degree, and ascorbic acid content decreased significantly. Among traditional products, *Mutchayan* looked like the most promising (Gbaguidi et al., 2022).

Marius Affonféré's research focused on the use of *Adansonia digitata* fruit pulp, *Moringa oleifera* leaf powder, and *Cochlospermum tinctorium* root powder to alleviate iron deficiency among children aged 6–59 months. Investigation showed that *Adansonia digitata, Moringa oleifera*, and *Cochlospermum spp.* are used by rural populations for many purposes. The *Adansonia digitata* fruit pulp, *Moringa oleifera* leaf powder, and *Cochlospermum spp.* root powder are the most commonly used from these plants. A complementary food supplement (CFS) was developed using these local food ingredients to alleviate iron deficiency among children aged 6–59 months. One-hundred grams of dry weight of the formulated complementary food supplement contained 17.4±1.1 mg of iron, 1.2±0.1 mg of zinc and 830.0±0.2 mg of calcium. The daily consumption of 10 g of CFS (dry basis) would cover 25.2%, 57.9%, and 42.4% of iron Estimate Average Requirements for children aged 6–12 months old, 12–36 months and, 36–59 months, respectively. The fermented maize and sorghum *ogi* porridges enriched with the CFS at substitution rates (in dry weight) of 15% and 16%, respectively, were found to be sensorily accepted by children and their mothers (Affonféré et al., 2021). Nevertheless, suggestions are made for future research on the mineral bioavailability of the developed CFS and the porridges enriched with this CFS.

Mariette Agbohessou drew on earlier research and worked with farmers and the colleges on baobab production. This study aimed at (i) characterizing the population structure, natural regeneration, and health status of baobab in the communes of Materi, Cobly, and Boukoumbé which are known as its hotspots in Benin; (ii) summarizing existing knowledge on the vegetative propagation methods of baobab and (iii) comparing performance of two grafting techniques to improve baobab leaf production: final slot grafting and side veneer grafting. From the grafting experiment, 50% of the grafted individuals succeeded, irrespective of the techniques. Side veneer grafting gave a rate of 85% against 30% for the final slot. In addition, from the grafting experimentations (Agbohessou et al., 2018a, 2018b), grafted seedlings have the advantage of having identical characteristics as the mother tree and show a shortened period of fruit production as they may start flowering less than five years after grafting instead of 8–12 years after plantation from seeds. Furthermore, previous results (Hounsou-Dindin et al., 2018) along with TVET students' research activities showed that planting from seeds for the production of baobab leaves in permanent gardens is also promising. For instance, young leaves of the species can be sustainably available as vegetables for consumption from 45 days after sowing the seeds (41.62 ± 1.16 kg of dried matter of leaves par 100 m^2). These results provided a solid basis for further domestication of African baobab with positive consequences for food security and nutrition and biodiversity as well.

The BAOCHAIN team generated a range of evidence to inform entrepreneurs and policymakers on the state of the baobab value chains as well as the opportunities and imperatives to boost the development of the market for baobab-based products. Six value chains of baobab pulp and leaves were identified, described, mapped, and analyzed.

Impact

This CARP project has impacted all the engaged institutions and the rural communities. The implementation approach (a multi-actor, action research innovation platform) was appreciated and seen as an approach to be promoted and taken up

(a)
(b)
(c)

FIGURE 21.1 Products ready for market developed from baobab pulp/leaves processing: (a) Baobab leaves ready for harvest; (b) Baobab products ready for market; (c) Baobab fortified infant food and baking flour

(a)

(b)

FIGURE 21.2 (a) Technologies developed for rapid baobab leaf vegetable production; (b) Side Veneer grafting and Final Slot grafting for quicker fruit production

more widely. This CARP initiative has contributed towards the organization of the baobab value chain actors on a national scale. This national platform was set up as a framework for dialogue between actors having a stake in the development of the baobab sector and sharing the ambition of its development. As participants in this CARP, 491 farmers were directly impacted, and more than 200 farmers were indirectly reached via our communication channels and also through visits to the baobab plantations installed in participating TVET colleges. They have improved their production capacity and their marketing of baobab products which has improved their livelihoods and their food security.

Eight Technologies, Innovations and Management Practices (TIMPs) on baobab have been developed in the frame of the project, and some of them have already been picked up by communities. The most widely taken up TIMP is the baobab leaves production. Two farmer associations (Coopérative Maraîchère de Cotonou and Eden Farm of Zè) in the southern part of Benin accounted for part of this take-up. Most of the supported students and interns are engaged in baobab-related products enterprises. Some of them have also been awarded fellowships to strengthen their businesses.[4] Participating in the BAOCHAIN has enabled the students to meet and engage with colleagues and experts from across Africa. It has opened up their understanding of rural communities and has strengthened their capability for business, further study, and effective employment as well as enriching their lives.

This project allowed the collaboration of researchers from two national universities (University of Abomey-Calavi and National University of Agriculture). Three laboratories of the Faculty of Agricultural Sciences (Laboratory of Applied Ecology, Laboratory of Biomathematics and Forest Estimations, and Laboratory of Food Sciences) were involved while two schools of UNA were engaged (*Ecole des Sciences et Techniques de Conservation et de la Transformation des Produits Agricoles* and *Ecole de Foresterie Tropicale*). This constitutes a multi-disciplinary team which is useful for future research initiatives within and between the universities. This will help to enhance future baobab value chain development and strengthen relationships to explore other areas of collaboration between faculties in the two universities. Working collaboratively on innovation platforms and the success of the baobab project has raised the profile of the universities and has triggered interest and investment to support more engagement in the future.

The experience with the engagement of TVET institutions was very beneficial. It shows that when universities and TVETs institutions come together to address agricultural development issues, there is faster uptake of the innovation, and such collaboration needs to be nurtured and strengthened. This experience has demonstrated the potential for closer collaboration between the colleges and the universities.

The upgrading of baobab pulp processing generated more wealth (added value of the activity between XOF 2,150 and 7,500 per kg of baobab pulp). Approximately, 400[5] out of the 700 farmers exposed to the opportunities have taken up the developed technologies. Furthermore, there has been a positive impact on consumers who now have access to more healthy foods and drinks. For instance, the baobab-fortified complementary foods are reaching many households with porridge, drinks, and baby food and contributing to malnutrition reduction. The dried baobab leaves have increased access in urban areas as well as rural areas to these nutritious vegetables. Growing them in farm gardens has also reduced the frequency of accidents due to the dangers of climbing trees to harvest leaves from the wild. However, the extent of these changes and the impact of the baobab CARP (BAOCHAIN) on the lives, nutrition, and livelihoods of different demographics needs to be measured in future research. This will provide valuable information to inform policy, not only in Benin, but across Africa in those areas where baobabs occur.

From the implementation approach of this project, the link between university actors and non-university actors (TVET institutions, NGOs, farmers organizations, and private sector) has been strengthened. Also, the engagements showed that formal scientists still need to learn from farmers and participating in this platform highlighted the importance of this engagement.

The CARP made it possible for much closer links between the researchers and collaboration with non-university institutions. It was useful to discuss, research, and engage in business with non-university staff members and with TVET students. Moreover, this project was a good opportunity for students since it allowed them to realize that it is possible to derive a business from their research in addition to producing a dissertation and broadening their network of contacts and their personal skills.

Conclusion

The implementation of the project has allowed the CARP coordinating team in general to improve their skills in collaborative research-for-action initiatives. It allows the faculty of agricultural sciences and the university to engage with other university and non-university actors. As key lessons emerge out of the implementation process, we could stress two: (i) When universities and TVETs institutions come together to address agricultural development issues, there is a quick uptake of the innovation and such collaboration needs to be nurtured and strengthened; (ii) There is important knowledge that formal scientists still need to learn from farmers.

Eight TIMPs on baobab were developed in the framework of the project and some of them have already been picked up by communities. There is potential to scale out the most promising technologies within Benin and across Africa to support biodiversity and alleviate poverty and malnutrition.

At the end of this project, we recommend that:

- Actors on the VC of baobab maintain their engagement and strengthen the national platform by developing joint projects and reaching out to potential partners
- National Government through extension services support such collaborative, multiple-stakeholder, multi-disciplinary, research-for-action initiatives.

Areas for further research

From this multi-actor participatory action research initiative, it is clearly noted that some aspects still need to be investigated in order to strengthen the value chain of baobab to facilitate its cultivation to improve biodiversity and its processing and marketing for better nutrition and livelihoods. For instance, future research actions are required to:

- evaluate the impact of the CARP on the farmers' livelihoods, on nutrition of consumers, and on wild baobab;
- evaluate impacts of livestock grazing on natural regeneration of the species;
- improve the economic and financial returns of baobab leaves production;
- assess the mineral bioavailability of the developed baobab-derived products;
- assess the impact that participating in the baobab value chain has on rural incomes;
- assess the level of uptake, and the likely nutrition impact, of baobab products by randomly sampling households and estimating consumption and impact;
- assess the potential of other baobab products such as the oil for local and international human health and beauty markets, and a range of other locally important baobab uses including livestock feed.

Interested readers are referred to a film on the RUFORUM Impact Platform https://ruforum.org/impact/project/a-formidable-baobab-research-team-ruforum-expands-support-in-west-africa/.

Acknowledgements

This platform would not have been possible without the active support of the farmers and entrepreneurs and our collaborators in Lycée Agricole Médji de Sékou and Lycée Technique Agricole de Natitingou; CIDEV ONG, Foundation UAC Start-Up Valley, ONG Potal men, ONG Jura Afrique, ONG Atacora Essentiels; Charl-Flor Technologies, Eden Business Farm, Atacora Essentiel, Coopérative Maraichère Cotonou and co-researchers Professor Bruno Djossa and Professor Belarmain Fandohan (National Agricultural University of Benin) and Dr Yann Madodé, Dr Valère Salako, and Dr Kourouma Koura (University of Abomey Calavi) and the BSc special project researchers.

Notes

1 The current project ends in March 2023 but the platform of engagement of universities and stakeholders continues.

2 https://youtu.be/mDCAnhO037w.

3 Mutchayan is a product derived from the fermentation, for 1 to 7 days, of the cooked cereal (maize, sorghum, or millet) dough mixed with baobab pulp.

4 Tony Elumelu fellowship to Mariette Agbohessou for enhancement of her business on baobab leaves production. Tony Elumelu fellowship to Mechak Gbaguidi for his business on baobab pulp production. Innovations, Products to Markets Field Attachment Program Award (IPM-FAPA) to: Kevin Fassinou for his business "Improving nutrition with a complementary food supplement"; Vanessa Idohou for "Baobab leaf marketing in the sudano-guinean zone of Benin"; Mariette Agbohessou for her business on baobab

leaves production (BAO LEAVES); Brigitte Sènakpon Assogba for "Akpan EKOFA du Roi", a traditional yoghurt enriched with baobab fruit pulp; Oluwakêmi Lysias Gael Houeto for "'BCMix' Corn flour enriched with the baobab fruit pulp".

5 The exact number is estimated from engagements on the ground but has not yet been rigorously assessed.

Bibliography

Affonféré, M., Chadaré, F. J., Fassinou, F. T. K., Talsma, E. F., Linnemann, A. R., and Azokpota, P. 2021a. 'A Complementary food supplement from local food ingredients to enhance iron intake among children aged 6–59 months in Benin'. *Food Science & Nutrition*, 9: 3824–3835. DOI: https://doi.org/10.1002/fsn3.2358.

Affonféré, M., Chadaré, F. J., Fassinou, F. T. K., Linnemann, A. R., and Duodu, K. G. 2021b. 'In-vitro digestibility methods and factors affecting minerals bioavailability: A review'. *Food Reviews International*, 39(2): 1–30. DOI: https://doi.org/10.1080/87559129.2021.1928692.

Agbohessou, M., Salako, K. V., Idohou, R., Gbèdomon, R. C., Hounkpèvi, A., Chadaré, F. J., Glèlè Kakaï, R., and Assogbadjo, A. E. 2020. 'Status of vegetative propagation of baobab: A review'. *African Crop Science Journal*, 28(s1): 215–224. DOI: 10.4314/acsj.v28i1.16S.

Agbohessou, M., Salako, K. V., Idohou, R., Hounkpèvi, A., Gbèdomon, R. C., Senon Victorin, Glèlè Kakaï, R., and Assogbadjo, A. E. 2018a. 'Fiche technique pour le greffage en fente latérale du baobab africain (*Adansonia digitata* L.)'. Laboratoire de Biomathématiques et d'Estimations Forestières (LABEF), Université d'Abomey-Calavi (UAC), *Série Agroforestière* N°: AGF/002/2018, Abomey-Calavi, Bénin.

Agbohessou, M., Salako, K. V., Idohou, R., Houénon, H., Hounkpèvi, A., Gbèdomon, R. C., Senon Victorin, Glèlè Kakaï, R., and Assogbadjo, A. E. 2018b. 'Fiche technique pour le greffage en fente terminale du baobab africain (*Adansonia digitata* L.)'. Laboratoire de Biomathématiques et d'Estimations Forestières (LABEF), Université d'Abomey-Calavi (UAC), *Série Agroforestière* N°: AGF/003/2018, Abomey-Calavi, Bénin.

Assogbadjo, A. E., Glèlè Kakaï, R., Chadaré, F. J., Thomson, L., Kyndt, T., Sinsin, B., and Van Damme, P. 2008. 'Folk classification, perception, and preferences of baobab products in West Africa: Consequences for species conservation and improvement'. *Economic Botany*, 62: 74–84.

Assogbadjo, A. E., Sinsin, B., Codjia J. T. C., and van Damme, P. 2005. 'Ecological diversity and pulp, seed and kernel production of the baobab (*Adansonia digitata*) in Benin'. *Belgian Journal of Botany*, 138(1): 47–56.

Chadaré, F. J., Hooiveld, G., Linnemann, A. R., out, M., and Hounhouigan, D. J. 2014. 'Effect of cooking on in vitro solubility of minerals and carotenoids in *Adansonia digitata* leaves'. *Annals of Agricultural Sciences*, 18: 1–19.

Chadaré, F. J., Linnemann, A. R., Hounhouigan, J. D., Nout, M. J. R., and van Boekel, M. A. J. S. 2009. 'Baobab food products: A review on their composition and nutritional value'. Critical *Reviews in Food Science and Nutrition*, 49: 254–274.

De Caluwé, E., Halamová, K., and Van Damme, P. 2010. '*Adansonia digitata* L. – A review of traditional uses, phytochemistry and pharmacology'. *Africa Focus*, 23: 11–51.

Gbaguidi A. M., Chadaré, F. J., Madodé, Y. E., Akogou, F. U. G., Assogbadjo, A. E., and Akissoé, N. H. 2022. 'Characteristics of baobab pulp-based foods from urban areas of Benin'. *Sciences and Technologies for Sustainable Agriculture*, 2(1): 36–45.

Gbaguidi, A. M., Chadaré, F. J., Salako, V. K., Idohou, Y. O. V., and Assogbadjo, A. E. 2020. 'Optimisation of oven-drying of baobab leaves using a central composite design'. *African Journal of Crop Science*, 20(1): 15–26.

Gebauer, J., Adam, Y. O., Sanchez, A. C., et al. 2016. 'Africa's wooden elephant: the baobab tree (*Adansonia digitata* L.) in Sudan and Kenya: A review'. *Genetic Resources and Crop Evolution*, 63(3): 377–399. DOI: 10.1007/s10722-015-0360–1.

Hounsou-Dindin, G., Salako, V. K., Idohou, R., Sero, N., Gbèdomon, C., Chadaré, F. J., Glèlè Kakaï, R., and Assogbadjo, A. E. 2018. 'Data sheet for baobab (*Adansonia digitata* L.) leaves' production'. Laboratoire de Biomathématiques et d'Estimations Forestières (LABEF), Université d'Abomey-Calavi (UAC), *Agroforestry Series* N°: AGF/001/2018, Abomey-Calavi, Bénin.

Hounsou-Dindin, G., Assogbadjo, A. E., Idohou, R., Salako, V. K., and Glèlè Kakaï, R. 2016. 'Morphological variation of baobab fruits and seeds traits in smallholders farming systems in Benin: A preliminary study on baobab leaves production'. RUFORUM Working Document Series, 14(2): 979–985. ISSN 1607–9345. 14(2): 613–620.

Idohou, V. Y. O., Egue, F. O. M., Madode, Y. E., and Chadaré, F. J. 2018. 'Effect of different pasteurization schemes on baobab nectar quality'. RUFORUM Working Document Series, 17(1): 219–226.

Institut National de la statistique et de l'analyse économique (INSAE) [Bénin] & Macro International Inc. 2007. 'Enquête démographique et de santé (EDSB-III)-Bénin 2006'. Institut National de la statistique et de l'analyse économique Bénin et Macro International Inc., Claverton, Maryland, USA, 492 p.

AFTERWORD

Growing from competitive projects to collaborative research programmes

Anthony Egeru

This book has documented the success and difficulties of collaborative engagement through university community action research platforms (CARPs) to address different smallholder commodity value chains across Africa. The projects have provided windows to see how transformative universities can be when they position themselves effectively as scaling laboratories and fulcrums of development. The projects have resulted in more proactive, engaged, creative and employable graduates. They have established relationships across disciplines and education institutions. They have made significant differences to the communities they have worked with and have helped to link a wide range of stakeholders together to increase agricultural productivity, rural livelihoods and climate change adaptation. They have raised the profiles of the universities and of the individual researchers.

Yet further to these accomplishments, the varied and diverse experiences of the CARPs, taken as a whole, can provide wider learnings to help shape new strategic directions for African universities looking towards the future. As targeted investments, the CARPs demonstrated that universities can effectively deliver appropriate science solutions for development – albeit within the limitations of project funding and allotted time. These initiatives allowed for powerful processes of learning, competence building and institutional alignment to occur. At the same time, the CARP experiences highlighted an underlying necessity to reframe how research and innovation is initiated within and across universities. Stronger and more comprehensive programmatic support structures are now needed to enable universities to mobilise their resources towards sustaining more cohesive research strategies. Working at a project level rather than an institutional level, the CARP approach places too heavy a burden on academics to mobilise resources when, in fact, there is a need to focus on facilitating the platforms of engagement that produce both the high level and the locally relevant research and graduates that are needed to drive Africa's resurgence.

Informed by evidence from the CARPs and graduate research grants, RUFORUM has developed a broader research framework that builds on the accomplishments of these prior initiatives (which will also continue) yet provides a stronger mechanism to enable institutions to focus on sustaining cohesive and systemic research strategies. The CARPs have demonstrated the effectiveness of the multi-stakeholder innovation platform approach in stimulating both higher education and agricultural transformation. With a new framework to shift the focus of research from project-based to programmatic, universities will gain leverage to mobilise their efforts towards sustaining and scaling out the multi-stakeholder platforms, amplifying the qualities of collaboration, co-creation, innovation and learning that they nurture across systems.

Building on robust cultures of collaboration, co-creation, innovation and learning, universities can demonstrate their relevance and sustain the quality and value of the education, research and positive impacts they offer in communities and in the wider economies of Africa. Going forward, RUFORUM will work with members for systemic change in their universities to better deliver the 21st-century skills necessary for critical thinking and reflection, adaptive and lifelong learning, active citizenship and future leadership roles. They will engage with members to rethink, rewire and re-envision learning policies with more focus on experiential and problem-based learning and 4IR skills. RUFORUM Research Programmes will focus on strategic dimensions and commodities that feed Africa and advance Africa's competitiveness, harnessing indigenous and local knowledge as well as 4IR technology. Within the Research Programmes, three clusters of universities within the RUFORUM network will work collaboratively; the well-established universities, mature universities and emerging universities. By aligning the universities this way, institutional capacity strengthening will proceed while resource leverage and the dynamic collaborative network will remain vibrant in the continent.

Building on the linkages created through the CARPs, RUFORUM has also deepened engagement with TVETs and is piloting a strategy for the colleges, with the assistance of the universities, to take a leading role in bridging the gap between farmers, new technologies and information through Community Based Agricultural Assistants (CBAAs). Drawing on the business and entrepreneurial successes of CARP students and communities, RUFORUM will encourage member universities to significantly ramp up their innovation hubs. They must identify key stakeholders to bring together to mentor, incubate and accelerate the growth of startups. They need to work with government and the private sector to improve the ecosystem so that it is easier to register and formalise trademarks and companies and to mentor and improve the skills of the youth.

The success of the graduates engaged in the CARPs has shown what can be achieved. Above all, universities must nurture human capital development which is the most essential investment for Africa. This is the strategic pillar which underpins the universities' continued viability. In the face of food insecurity, climate change, poverty and other complex challenges, educating people innovatively is the golden key that will unlock Africa's profound progress into the future.

INDEX

Note: Page numbers followed by "n" denote endnotes.

For Product Safety Concerns and Information please contact our
EU representative GPSR@taylorandfrancis.com Taylor & Francis
Verlag GmbH, Kaufingerstraße 24, 80331 München, Germany